CREATIVE BURST

A PRACTICAL METHOD FOR UX DESIGN TEAMS TO DRIVE INNOVATION PRODUCT STRATEGY FROM WITHIN

KRISTIN MORIN

INSIGHTFUL SCRIBBLES

ISBN: 978-1-64355-414-3

For Dory
The one who made my Creative Burst journey, including this book, possible.

CONTENTS

INTRODUCTION

Organizational recognition and authority—yes, please! User Experience (UX) practitioners are too familiar with the constant struggle for recognition and authority if they work in environments that haven't yet fully embraced a user-centered design model. True organizational change takes years. But in the meantime, ambitious UX teams can drive creative leadership.

This book presents the case study of the first UX-driven innovation program we started at Kronos and how we found ways to transform team insights into product vision that impacted the future of our enterprise software solution.

Yes, product managers owned responsibility for defining roadmaps and product direction. Yes, legacy systems bogged down grand visions for user-focused features. Yes, employees outside the team often perceived UX activities as "look and feel," and leadership often told us, "we don't have time for research."

Within those all-too-familiar constraints, we established an outlet —Creative Bursts—that let us explore a diverse set of important problems and watch for the right moments to share our visions.

Some ideas fell by the wayside; some ideas turned into larger projects with formal budgets.

While one key difference relates to the strength of the idea, much of it amounted to luck. By "luck," of course, I mean the convergence of preparation and opportunity.

Our Creative Bursts prepared us for the moment the opportunities came along—the UX team determined a vision, and we positioned ourselves to lead with it.

This book will offer a case study of our program, how the program evolved over its lifespan, and how we poised ourselves to leverage our insights into product vision.

You'll also witness the ups and downs of my personal journey through my 18 months of facilitating our Creative Bursts. You'll witness how the moment of serendipity when I joined the first planning meeting, as well as the hard work and experimentation that followed, changed my entire perspective. The perspective on both my UX career trajectory and my definition of what it means to work in a great software company.

I can't share every anecdote or project story I'd like to. You won't see workshop photos or real design artifacts. I am obligated to withhold intellectual property, so I'll address in broad strokes how our explorations impacted strategic projects. But I'll dive into as many specific stories as I can, create exemplar visuals just for this book, and share my personal experiences and insights.

After reading, you will come away with a framework of building blocks for creating your own UX-driven innovation practice—either formal or informal. You'll also understand the toolkit you need to leverage these building blocks in whatever way makes sense within your organization.

But mostly, I hope you'll come away with the excitement and confidence to tackle these challenges yourself. Because anyone can. You only need to embrace them.

WHAT TO EXPECT

The book's structure consists of three parts:

- Definitions and perspectives
- Creative Burst case study
- Creative Burst takeaways & general principles

Definitions and perspectives

This will be a quick chapter to set the context for the case study. Who are you? Will you benefit from my perspective? Why should you consider an innovation practice at all?

Creative Burst case study

We'll begin with a quick overview of the environment at Kronos that produced Creative Bursts, so you can determine how strongly it parallels your own situation. Or doesn't. If your organization is more ready or less ready for UX-led design innovation strategy, you'll need to adapt your approach accordingly (and possibly, your level of secrecy).

Then we move into the case study, which will be the "meat and potatoes" of the book. I'll provide an overview of our original concept for Creative Bursts, explain how and why we iterated the program, and share some examples of the successes we achieved.

Creative Burst takeaways & general principles

We'll wrap up as I reflect on my experience and analyze the key takeaways from Creative Bursts and principles that yielded success and failure at the respective stages. I'll discuss foundations and three building blocks—problem identification, solutions design, and opportunity visibility—as well as skills and strategies for success at each stage in the process. Finally, I'll present an example sequence of Creative Bursts to show you how to put it all into practice. As a bonus, I'll provide a roundup of helpful and inspirational resources to dig deeper into these topics as you begin your journey.

Welcome along. I hope you enjoy the ride.

PART ONE

DEFINITIONS AND PERSPECTIVES

CHAPTER ONE
INNOVATION WHO, WHAT, WHY, AND HUH?

The rhetoric of innovation is often about fun and creativity, but the reality is that innovation can be very taxing and uncomfortable.

—*Linda A. Hill, Collective Genius*

WHAT IS INNOVATION?

BEFORE I PROCEED, we need to discuss the term "innovation." After all, you'll struggle to benefit from the "how" of this case study without a fundamental understanding of the "what" you're trying to achieve.

Innovation seems to be the unicorn of large software/web organizations in the 21st century. A mythical creature. Everyone serious about growing a futureproof business knows they need to "innovate" to stay relevant while newer, leaner, more specialized businesses pop up around them.

But what does innovation actually mean? Is it just jargon—so overused we can't possibly find a meaning that we can put to immediate practical use?

Well, yes.

And no.

Just because a word has been diluted by overuse, misuse, abuse doesn't mean we can't refocus on its essence.

Anytime you're struggling with professional jargon, I encourage you to shift your mindset. While business language carries a lot of baggage for many of us, it's also one of our best tools. It allows us as UX teams to convey to our organizations what we can offer.

Embrace all that what the language of business is loaded with and use it to your advantage.

I could define a new, baggage-free vocabulary, and I'd be signing you up to learn that vocabulary, then teach it to your company. A book that purports to be a "practical guide" shouldn't demand that kind of burden. We've made such strides teaching our organizations the value of UX design over the past few decades; isn't it time we leverage our past successes rather than biting off an entirely new education task?

Despite the confusion, ambiguity, and inconsistency involved in the software/web industry's use of the term "innovation," its potential is too important for UX teams to simply shrug and move on. Here's my attempt to establish *some* kind of definition that will create the underlying frame of reference for this book, even if only in the context of what I'm trying to encourage.

> *A successful innovation practice empowers employees of an organization at all levels to identify unmet customer needs, then leverage that knowledge: to envision, to design, to build, and to test features, products, and services that align with an organization's goals and complement existing products and services.*

You'll ultimately achieve that outcome using the toolkit provided by the three building blocks I'll discuss later.

Problem identification uncovers unmet customer needs.

Solution design skills enable envisioning, designing, prototyping, and testing features, products, and services. (For most teams, "build" probably comes later.)

Increase opportunity visibility by showing how our solutions fit alongside our organization's current products and services.

For now, I'll leave it at that. I'll unpack the nuances of that definition throughout this book.

WHY THIS BOOK?

At this point, you're probably wondering who this person you've never heard of is, and why she's qualified to write a book on innovation.

Let me assure you: I don't consider myself qualified to write a broadly applicable book on innovation, nor do I intend to write any such book. But I helped develop—and subsequently facilitated—the first structured innovation program within the Kronos UX Design team. And I do state with all confidence that I'm qualified to tell my story and share what I learned from it. That's all I intend to do.

Realization and knowledge you need innovation to drive sustainable success is one step; getting your ideas heard is the next.

Here's the good news: if your team hires smart, capable workers who thrive in their current jobs and care about your company, you already have what you need. In fact, your team should be eager to contribute in more meaningful ways and have your ideas heard.

But to build a practice within your organization, you need a system, especially if your organization currently lacks a design-driven mentality.

This book provides a case study of one such system. Along the way, I'll share anecdotes, tips, and references to some of the key reading material that inspired me.

With luck, you'll find my case study relevant to your own situation and embrace its guidance on how you might create a similar practice within your own organization. If not, return this book. I promise I won't take offence!

WHO ARE YOU?

Primary roles

While I certainly don't claim to be able to tell you who *you* are, I'll share some assumptions on who you might be. Notably, I will spend the duration of this book speaking to you as if you're a UX leader or team member.

UX design leadership

You've got a team of eager, capable designers, but your organization is possibly not as far along the UX maturity model as you'd like, and you're still on your journey to earn UX a proper seat at the proverbial table. You feel like giving the team a bit of freedom. They should pursue their own ideas, as long as those ideas align with company goals. This is a great way to achieve two valuable strategic results: you will simultaneously instill job satisfaction in your team members and illustrate to executive leadership that your team could provide a crucial added value.

UX designers

You certainly can't deliver innovation without the insights and support of your colleagues, but you'll be the one to bring it all together. You'll be the facilitator. You'll be the one applying design thinking methods to problems. And because you're already a curious, inquisitive designer, you've cultivated a holistic view of the product and service suite, so you see connections that other roles might miss because of how intensely they focus on their current projects.

UX researchers

You learn about unmet customer needs every day. While many insights immediately impact active design and development work, you always uncover additional problems that irked end users. Those problems are a gold mine of opportunity. You're looking for guidance on how to partner with your design and development colleagues to transform those needs into something you can validate with users. That means designs, or even functional prototypes.

Type of organization

Next fundamental assumption: you're considering creating, or you're struggling to create, an innovation practice within a software company. Mainly because I'm familiar with how software gets created and can address where in the process these activities fit well.

For everyone else

I have no doubt there are folks in plenty of roles I've never even considered who are eager to impact innovation within their organizations. After all, the very nature of encouraging grassroots innovation means that I wish for employees at all levels of an organization to embrace and drive the process—and within all departments.

So if I've missed you, please feel free to reach out and let me know who you are and why you sought out this book. I'd love to hear about your perspective, your favorite takeaways, your triumphs and your failures.

WHY *NOT* AN INNOVATION PRACTICE?

I should be up front that even a well-planned system for innovation won't work in an organization that isn't ready for it, or one that isn't in full control over the work it does.

Startups: If you're working in a startup, you're probably fully focused on bringing a single innovation to life. Diverting attention to other work would truly jeopardize the company's ability to deliver the value it was created for.

Low UX maturity: If you're working in an organization that hasn't embraced design thinking methods, or your team is too small for multiple designers and researchers, you'll probably struggle to implement a lot of the mechanics without executive support. Or you'll end up working extra hours, which I never advocate. So if you still can't get a designer assigned to every project, or if you can only get research budget for prominent projects, you probably need to spend your energy on moving your organization further along the UX maturity model. Help create the environment that will embrace innovation, and work to instill innovation principles within team processes as you go.

Toxic environments: If your company is highly political, has seemingly unbreakable silos, or is another type of toxic environment, an innovation practice is likely not the right move. The company's leadership won't be receptive, collaboration will be exceptionally challenging, and you'll experience more frustration than anything else. Please, don't sign yourself up for that.

Consultants: If you're working in an agency or consultancy, this method is probably too simplistic to fit your needs well. You, and your company, might even already be skilled at designing innovative solutions. But you work for clients. You need to sell extra time and budget for innovation design and need to instill a tolerance for the necessary experimentation and ambiguity. That's no small endeavor. Over time, you'll spend considerable effort to convince a few clients to go along for the ride. It will take longer to produce a track record of success—and the resulting portfolio—to build up the kind of reputation that gets you invited to compete for these kinds of projects.

While I have no doubt you could learn something from the general principles contained in my case study, you won't find any information here tailored to helping your kind of company accomplish this objective. I haven't done it personally, so I can't share insight on the process.

IS YOUR COMPANY READY?

Even if the aforementioned examples don't describe your company, you still might be wondering if you work in the right kind of environment. I can't really give you concrete heuristics to evaluate whether your company *is* ready. Here are some signals that your organization *might* be ready.

1. You've heard the executive leadership calling for

"innovation" in vague terms or acknowledging that great ideas come from everywhere. (For any executives among you, I hope to forestall that vagueness with this book.)

2. Designers are allowed, even encouraged, to participate in other designers' projects. Diverse team members join for sketching sessions during exploratory design phases. Research sessions invite anyone to act as a volunteer note-taker, or simply spectate, for those who want to learn more about customers.
3. Researchers or customer support team members enter customer insights into a centralized database that employees can access. An even stronger signal is if you regularly mine the data for overlooked insights or apply machine learning to it.
4. Product owners explore nascent concepts with customers, supported by rough designs or prototypes, to determine if they are feasible to add into a future roadmap.
5. All employees are expected to fail as a learning tool to get to the right solution faster.

These signals, and many like them, suggest you have an environment that doesn't scrutinize employees to make sure they're maximizing every work hour. An environment that puts responding to customer needs ahead of operational efficiency.

Not every program will be as generous as Google's famous "20% time," but employees need space and time to learn, collaborate, think, and work on innovative solutions.

Innovation comes from trial and error, and that's not efficient. But it can be effective.

WHY AN INNOVATION PRACTICE?

Finally, right?

I imagine that the fact that you've found this book means you already want to create a successful innovation practice, or simply want to see how you might be a more innovative individual within your organization.

So you've survived the minefield of my caveats and warnings, and you still believe your organization might just be the kind of fertile soil in which you can sow the seeds of innovation. Now you might find yourself needing some arguments on why you and your team won't be "wasting" the time spent on innovative thinking.

I certainly can't do justice to the extensive works of Clayton Christensen, the father of modern innovation theory, but the justification you need lies in disruption theory.

The general idea is that solutions exist somewhere along a disruption curve. It may take years, or even decades, but an innovation like automobiles renders a horse and buggy obsolete when the population eventually embraces it. Long distance service becomes antiquated when everyone carries a smart phone. Stubborn people resist new technology entirely (to understand *that*, look up adoption curves), but most people will eventually stop using a typewriter in favor of a computer.

I highly recommend finding one of Clayton Christensen's articles[1] as a 10-minute introduction. For a deeper dive, jump right to his seminal book, *The Innovator's Dilemma.*[2]

Your innovation practice can—and likely will—spend much of its attention on incremental innovation. In the end, these ideas will result in continuous improvement to your organization's products and services. However, the improvements will be more effective and better informed than those your organization would experience if purely directed top-down. A bold claim, yes, but an inevitable consequence when the people who interact with customers every day drive improvements.

Radical innovation—and the potential for your company to be disrupted by it—is what likely keeps software executive leadership up at night. With an entire organization watching your marketplace, responding to customer needs, and identifying unmet opportunities and new ways of working, your company increases the chance of being its own disruptor. Outside disruption is inevitable. But when this external disruption finally renders your original product or service obsolete, pursuing and implementing radical innovations that serve your own customers should offer a safety net.

PART TWO

CREATIVE BURST CASE STUDY

CHAPTER TWO
DESIGN ENVIRONMENT

KRONOS IN TRANSITION

It is not the strongest of the species that survives, nor the most intelligent; it is the one most adaptable to change.
—Charles Darwin

I ARRIVED at Kronos as a contract-based Interaction Designer in the fall of 2014.

At this time, the company led the Workforce Management industry with a 15-year-old enterprise software solution called Workforce Central and had been recently recognized as one of the top employers in Massachusetts. They were in the process of transitioning from an on-premise, installed software solution to a cloud-based, Software as a Service (SaaS) solution and knew that the legacy baggage from that 15-year-old system was starting to hold them back.

So the design team supported the initial stages of a massive redesign of the entire enterprise software suite; with each passing month more designers transitioned away from the legacy software to the new solution.

At the winter company meeting, the CEO told the story of the project that initiated the company-wide effort. He formed a small team and challenged them to "figure out how to disrupt Kronos."

That skunkworks team spent a year in an offsite office. When they returned, they presented a grand vision for a "reimagined" version of the enterprise suite, claiming it would require five years to execute. They were told they had three years. (It eventually did take around five.)

Having made a hobbyist study of innovation and disruption, which included an elective in my elite business graduate school, I could make out the red flags of this story in abundance. This wasn't "disruption," it was incremental innovation. No matter how much improved the solution would eventually be, ultimately the vision was the same solution, but better. Or from an innovator's point of view: better, but still the same.

I don't intend this critique as a criticism of Kronos's efforts. The monumental undertaking was impressive to behold and thrilling to be a part of. No, I mention it to illustrate the difference between a top-down innovation mandate and the grassroots effort we eventually created. Without the context of "where they were" when I joined the company, you can't appreciate the transformation that allowed the eventual creation of Creative Bursts.

When Kronos rallied around the five-year effort to redesign and rebuild what would eventually be Workforce Dimensions, the entire company began to embrace the potential for true innovation.

UX DESIGN AT KRONOS

Achieving great UX design is not just a function or talent of individuals, it is an organizational characteristic.
—Chapman & Plewes

The UX Design team at Kronos operated as a centralized design agency within the larger product engineering division. The head of the UX team, who became the first VP of UX during my tenure at Kronos, also managed the product owners and the technical publications team (responsible for product documentation). Her broader focus infused our team with a predisposition for close cooperation.

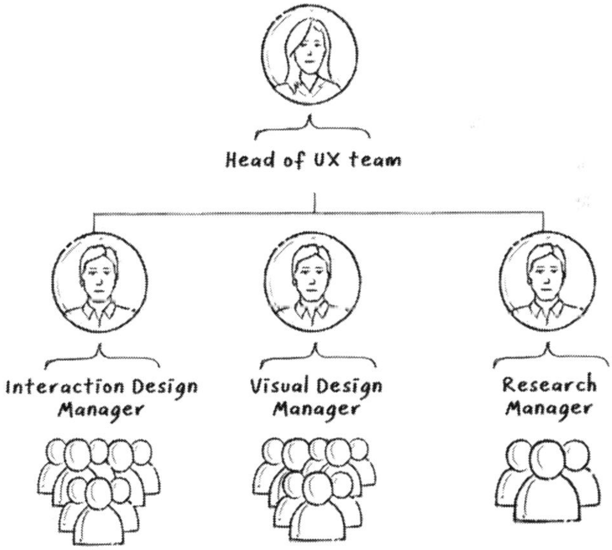

Team structure

Three disciplines comprised the UX team: Interaction Design, Visual Design, and Research, with team members distributed between our Massachusetts and Montreal offices.

The somewhat misleading title of Interaction Designer really meant "all UX tasks that weren't visual design and research." The 8–10 of us within the Interaction Design discipline led the discovery process, facilitated sessions for brainstorming and design refinement, created information architectures, performed task analyses, pitched designs to stakeholders, documented behaviors, and eventually created our first design system, alongside the tasks more typically associated with interaction design.

Another 8–10 Visual Designers handled the straightforward, yet complex, effort of transforming wireframes into the actual look and feel. They handled fonts, colors, visual language, icon design, style guides, detailed design specifications, and eventually most of the heavy lifting for our accessibility implementation.

3–4 Researchers, and a Creative Technologist who was responsible for building prototypes for internal evaluations as well as customer testing, handled both formative and evaluative research. They collaborated with interaction designers and product owners to understand research goals and write interview scripts and test protocols, they conducted usability testing sessions, analyzed findings, and reported back to project design and development teams.

During my time at Kronos, I watched the research team's role develop. They spearheaded more formative research to uncover new opportunities and understand deeper customer needs rather than simply "testing to see if this design is working for customers." The research team's shift happened alongside our innovation efforts, and it was a powerful signal of a deepening customer-first mentality.

Upon my arrival, the UX team didn't yet have any formal definitions of, or responsibility for, activities associated with Content Strategy or UX writing (a relatively new term in the industry back then). My background with both eventually infused core principles into our daily processes. Individual team members rarely got involved with product design strategy or worked on projects not assigned by a manager in response to a request from a product owner.

Project structure

Every project had a product owner and an agile development team. Designers supported the agile process once designs moved into development—edge case design, design revisions to address technical concerns and budgetary changes, and design quality assurance.

At that point, we operated via a mostly waterfall method. Yes, the team had learned how to embrace a more agile methodology and collaborate with the developers. But design changes from the legacy product were all-encompassing. To achieve our company's vision, we required big discovery and big planning: "Big Design up front."

Many projects covered an entire product within the suite, but at times we focused on a single large feature or improvement due to the schedule when development teams would start their work. A project covered a single "headline," which was our term for a unit of design and functionality, regardless of how large it was.

The term "headline" gave us the freedom to keep using our internal vocabulary even after the launch of the wholesale product redesign, and the team would need to shift back to supporting smaller, continuous improvements, just as we'd shifted into the waterfall-like model.

These standard projects—the ones Kronos was already doing when I arrived—constitute what I'll refer to as "roadmap-driven projects" going forward, in contrast to "innovation projects." The bulk of our time was spent on the day-to-day work of designing known products with requirements that were determined for us before we got involved. We often needed deep discovery to fully understand nuances of our customers' needs, but we fundamentally grasped the underlying problem.

UX team maturity

Over my first couple of years at Kronos, which was the time leading up to Creative Bursts, the atmosphere within the UX team changed dramatically. I can't quite pinpoint what caused it, but it correlated with certain tangible shifts.

The most noteworthy one took place once the entire design team became 100% dedicated to the reimagined product suite. With the entire team working on the same product suite, we started assigning multiple designers to large headlines, and this creative collaboration naturally led to deeper design thinking.

Increased design visibility within the product division led to our team head becoming a VP, and the team leads for all three disciplines received a title of director. The obvious message: the company valued design, and they rewarded that with responsibility and ownership.

Within our Interaction Design discipline specifically, we frequently reviewed each other's in-progress work for consistency and provided constructive design feedback. We talked about emerging tools and industry best practices. We shared anecdotes of customer sessions we'd witnessed.

Each year, we identified lofty goals for improving ourselves as interaction designers. We created manifestos outlining progress we wanted from ourselves, our direct boss, and UX leadership as a whole. We iterated on process to determine where we worked well and where we could improve with specific initiatives. Practical initiatives like creating a formal procedure for design kick-offs helped us facilitate better communication and collaboration on projects from the start. Ambitious initiatives encouraged designers not just to deliver obvious solutions that would certainly be "good enough," but to *always* provide a concept that considered the problem differently or pushed boundaries—regardless of how likely product owners were to embrace it.

But more importantly, those innovation-inspired initiatives signaled to us that we were eager to do more. Without the struggle to prove ourselves as valuable *contributors* to the company, we poised ourselves to take the next step. We recognized an opportunity to establish ourselves as valuable *leaders*. This was our moment to seize it.

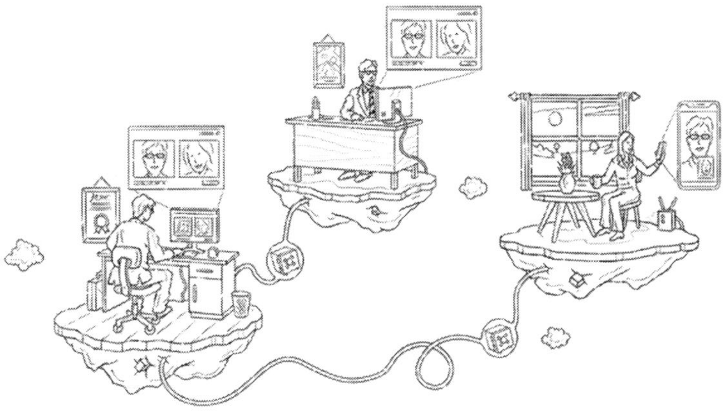

Physical location

In this post-COVID professional environment, I also should address the proverbial "elephant in the room" of physical location. This is not a book about remote vs. in-person collaboration. I will continue to speak in general terms about collaboration and "gathering in a room," but please understand that I don't necessarily mean that literally.

Our team was distributed across two offices, Massachusetts and Montreal, and while we aimed to participate in person at one of the two offices on Creative Burst days, we were constantly connected by video conferencing. Any team members who couldn't get to either office simply joined remotely. Even then, we had a generous "work from home" policy, and team members occasionally being remote was just a fact.

Certain specialized exercises would require an extreme amount of planning and preparation to accomplish with a fully distributed team—like the physical prototyping exercise on timeclocks that I'll share later in the case study. But I trust I'd find a way to run the program even without the benefit of physical proximity, were I to construct the program in a remote team environment. And you certainly can too, especially if you've already been collaborating as a successful distributed team.

CHAPTER THREE
CREATIVE BURST 1.0—L.E.A.P.S. FORWARD

TO THIS DAY, I'm still not entirely sure what earned me an invitation to that first planning meeting for the UX team-driven innovation program. Perhaps my habit of sharing non-fiction books on innovation and creativity. Perhaps my track record of advocating continuous design team improvement. Perhaps my boss recommended my inclusion.

Perhaps they were simply planning ahead, knowing they'd need to delegate day-to-day logistics. I don't think I'll ever really know.

Yet there I was in a room with the three team leads, brainstorming the structure for a program that would nurture designer-led product innovation at Kronos. Because I was the least busy of the four of us—being the only individual contributor among managers ensured that—I ended up overseeing the minutiae of the program we came up with. Plus, I love facilitation.

CREATIVE BURST STRUCTURE

> *If your goal is innovation, then your role must instead be to create an environment—a setting, a context, an organization—where people are willing and able to do the hard work of innovation themselves: to collaborate, learn through trial and error, and make integrated decisions.*
>
> *—Linda A. Hill, Collective Genius*

Spoiler alert, the first iteration of our Creative Bursts was needlessly complicated. But from analyzing how excessive complexity jeopardized our success, we learned a lot about what worked and what didn't, so I still think it's worth sharing. I just don't recommend using it as-is.

We envisioned the core of the program as a game, where individual designers or small teams would conceive of an idea and compete to have their vision realized. As the pool of ideas narrowed, designers would form bigger and bigger groups around a few ideas, ultimately leading to full team collaboration on a single idea that we'd pitch to executive leadership.

The framework was structured around the acronym L.E.A.P.S.—Listen, Explore, Act, Persist, Seize—signifying the leaps forward we'd enable. Each phase revolved around a central principle that drove the activities. Plus, we loved our clever acronyms.

The three UX team leads would act as both mentors and judges throughout the process—they formed the innovation committee. At least, that was the theory.

As I've mentioned, we recognized critical flaws quickly and we pivoted prior to completing our first full game, but I think the philosophical underpinnings are still valuable to understand. As is our attempt to scaffold a process for driving concepts from idea to solution to pitch.

Innovation time and space

You need to create space for your creative process to thrive rather than expect it to operate in the cracks of your frenetic schedule.

—Todd Henry, The Accidental Creative

For the first time in the UX Design team's history at Kronos, the logistics of the program provided a formal, scheduled time for innovation-focused design activities.

Frequency: biweekly on Thursdays
Creative Burst: 12:00 p.m.–1:00 p.m.
Exploration and execution: 1:00 p.m.–5:00 p.m.

Every two weeks on Thursdays, the entire team gathered at lunch for the Creative Burst game session (food provided by the company), followed by an afternoon of free collaboration and creativity time.

During that time, designers weren't expected to make progress on headlines, attend project meetings, or respond to emails. We were expected to be spontaneous, to jump into a conference room with teammates and collaborate on a whiteboard (physical or virtual), experiment with novel design techniques, embrace new technology, and drive design forward at Kronos.

The L.E.A.P.S. game played out during the one-hour lunch sessions. The ideation we needed to conceive and refine concepts took place during the afternoon collaboration sessions.

Team members submitted topics for us to evaluate the idea's

potential. These idea owners would continue to guide their conceptions through the process.

Across the five phases, the group would simultaneously develop and filter ideas at every stage as everyone voted for the strongest ideas to continue through the process.

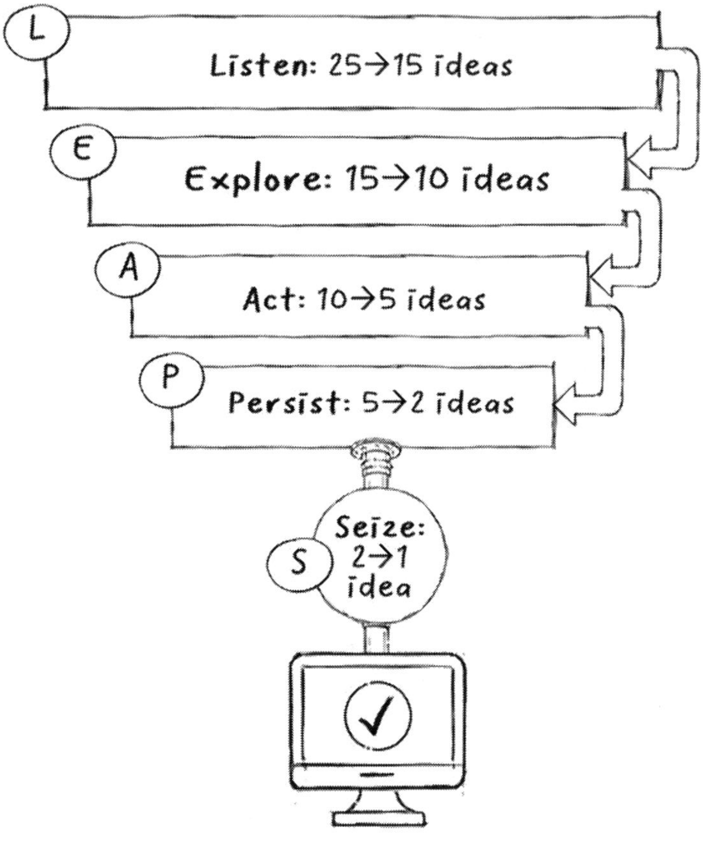

Our intended result: one full game session per year resulting in one stellar, well-vetted pitch.

Listen

Innovation: Imagine the future and fill in the gaps.
—Brian Halligan

Perhaps counterintuitively, this call to "listen" wasn't about learning from customers or seeking outside inspiration: it was about listening to yourself. Disruptive innovation leads the marketplace to something customers didn't know they needed or wanted, rather than following demands for a better version of what they already have.

Since our team already had an excellent culture of learning from customers, we wanted to challenge designers to trust their own instincts and ideas, knowing that their previous learning would steer them true.

During each of the five Listen Creative Bursts, idea owners described an idea or vision in a few sentences. They then presented a 2 ½-minute explanation, followed by another 2 ½ minutes answering audience questions, with no visuals allowed.

Explore

Exploration is curiosity put into action.
—Don Walsh

Exploration is a critical element for designers hoping to break free from comfort zones and embrace new ideas and new ways of working. To adopt new perspectives, this call to "explore" included the literal, changing physical location and working environment as well as the figurative, i.e., seeking out and sharing trends and inspirations from different industries—especially consumer-focused industries—with teammates.

During the explore cycles, idea owners and their collaborators would refine and expand their concepts. Together they would create meaningful visuals in whatever form best conveyed their ideas: wireframes, storyboards, sketches. Whatever illustrated the concept's key value for customers and for Kronos.

Act

Ideas are useless unless used. The proof of their value is their implementation. Until then they are in limbo.

—Theodore Levitt

The call to "act" was the workhorse concept of the program. As any venture capitalist will tell you, there's a reason you can't copyright or patent an idea. Ideas are cheap.

To truly understand how—or even if—they work, designers need to transform them into tangible artifacts like sketches, wireframes, prototypes, and storyboards. We expected designers to treat ideas as a flexible starting point, to take a series of small, simple steps. An iterative process of executing on ideas and learning from the results.

After validating the viability of the core concept, we challenged idea owners and their teams to detail the major use cases involved in executing their vision. Once again, they would package these use cases, and associated visuals, into 10-minute presentations that illustrated depth of thought and an understanding of the problem space.

Persist

Failure isn't a necessary evil. In fact, it isn't evil at all. It is a necessary consequence of doing something new.

—*Ed Catmull, Creativity, Inc.*

If "acting" was the workhorse of the program, persistence was the workhorse's guiding reins. Our experienced team already excelled at providing sound solutions to tangible problems and delivering designs that met requirements and received prompt approval.

But innovation demands more than simply pushing beyond design comfort zones, plus we needed the resilience to continue trying new ideas even though they might go nowhere. We encouraged designers to start without fear of failure, and to keep going. To carry on when they experienced unexpected shifts in direction or even resistance.

To. Keep. Going.

For the final five ideas, idea owners would present for 10 minutes every week. The innovation mentors and UX team members helped define what visuals would best convey critical elements of problems. Idea owners might need to focus on deeper exploration of the core concept, or they might need to consider and design around edge cases. Two of these five ideas would move on.

Seize

Opportunity is missed by most people because it is dressed in overalls and looks like work.

—*Thomas A. Edison*

At the end of a L.E.A.P., the call to "seize" required cooperation and commitment. As we evolved ideas from nascent concepts, and determined their potential for further development and research, many ideas would fall by the wayside. Once we'd determined which would move forward, this was our rallying cry for the entire team to support and contribute to the ideas of others.

Jeff Bezos is famous for pointedly asking team members with a dissenting opinion if they were willing to move forward and "be wrong with me," rather than continuing to openly challenge the chosen course of action once he had reached a decision. It was his way of acknowledging the value of the opinion or alternate solution they'd worked so hard to prepare and present, while gaining their commitment to make forward progress, together. That philosophy was the centerpiece of the ultimate step.

Idea owners would expand their teams yet again during the first three Creative Bursts of the Seize cycle. Together, idea teams would craft a coherent narrative around the idea, structure visuals to support that narrative, and create the pitch deck.

Theoretically, everyone on the UX team would choose to participate in one of the final projects.

The following week, idea owners would deliver their pitches to the innovation committee. Those three individuals would choose one of the two ideas to present to executive leadership.

We'd then start the process over again.

L.E.A.P.S. IN PRACTICE

As I've already stated, the game didn't even last an entire cycle. I think we only made it a couple sessions into the Explore cycle, which we even started early, before participation dropped to critical levels. While there were certainly successful elements we maintained when we pivoted, the glaring failures threatened to cannibalize the program from within if we didn't make a change.

Program introduction

The very first innovation-focused team meeting introduced the game, its goals, its rules, its structure.

Our Creative Director (the team lead for the Visual Design team) presented a few slides sharing the L.E.A.P.S framework, the logistics, the biweekly free afternoon for innovation activities, some inspirational quotes from prominent innovation advocates, and a sampling of opportunities we might explore—wearables, automation, Internet of Things, challenges of supporting work-life balance for our customers' employees, tools for the growing gig economy. Many of which team members already championed as personal causes within their existing work.

My read of the room following the introductory meeting: reserved excitement, with a heavy dose of hesitance. I'm not sure if people were skeptical about truly getting a free afternoon to explore their own ideas every other week, or if they simply exhibited the confusion that so often follows the vague rallying cry "now let's go innovate!"

Despite the lukewarm reaction, we'd introduced the new program, and now it was my job to facilitate the heck out of it. Before they would embrace the opportunity in front of them, I needed people excited.

Not just so we could eventually achieve our stated goal of increasing the visibility of our team's capabilities. But because the promise of guiding the process had sparked an enthusiasm I'd never realized I'd been missing in my UX career.

My world would be back to roadmap-driven projects and standard UX process if this fledgling program never even made it off the ground.

CREATIVE BURST SESSION: WORKFORCE PROBLEMS IDEATION

The second session was purely an ideation session to prepare people for the game. I knew from the last two years that many team members already dreamed of pet projects they'd work on "when they could." I myself had an unrealized idea inspired during my first few weeks learning about the Kronos product offerings that I'd been eager to delve into.

"When *I* could."

Goal

To create a sustainable program, we needed a pool of customer problems to entice the team long after everyone worked through their existing "I've had this idea forever" backlog. We didn't yet have any formal systems in place for generating problems, and at this point we rarely conducted formative research.

But I had several notebooks from my first couple years at Kronos. They contained high-level takeaways and observations from projects I'd been involved in, research sessions I'd attended, internal presentations on marketplace trends, and customer challenges I'd heard from various other teams.

I wanted to launch the discussion from a tangible starting point but still create the opportunity for the team to engage with a brainstorm, so I distilled my notes into a list of themes that represented core challenges large organizations tackle every day.

Themes like:

- Offering enticing benefits
- Building a strong company culture
- Appreciating employees
- Retaining talent
- Encouraging wellness
- Attracting boomerang employees
- Focusing on work-life balance

None of these alone would be tangible enough to design solutions for. But I aimed for something different: enough specificity to spark team members' memories of concrete problems they'd already encountered throughout their own tenure at Kronos.

Participants

The entire team, all 25–30 of us, including the three team managers.

Visuals

For this verbal brainstorm session, I didn't bother preparing slides. I had a spreadsheet of the dozen or more themes I'd cobbled together, shared this with all team members while we delved into our problem space. In retrospect, this kind of drilldown on high-level themes would lend itself beautifully to mind-mapping.

Session activity

Throughout that hour, we ideated, expanded, and refined problems.

Most interestingly, a deep insight emerged as a natural byproduct of this exercise: what team members cared about personally. Of the 25 themes, five remained untapped, while we expanded "focusing on work-life balance," already a hot workforce management topic in 2016, into 14 distinct problem areas.

In all, we ended up with an impressive 95 problems for 20 of the 25 themes, and I could already see the wheels turning for team members. Problems sparked ideas for solutions. The shroud of hesitation from the program introduction began to melt away, to give way to excitement and inspiration.

Next up: I needed to nurture that spark.

Follow-up exercise

Following the session, I posted the full list of both themes and problems to inspire ideas for the first round of the game. Our first attempt to forestall the loss of fleeting ideas and allow our team members to reference them at the right moment.

Outcome

This list of ideas not only provided the tangible benefit of idea starters, but represented the beginning of our Creative Burst wiki, which would eventually drive the entire program as a digital home.

Hindsight adaptation

To help our visually inclined team members process and respond to the connected nature of our themes and problems, if I were to do this again, I'd choose a more visually oriented presentation like a mind map.

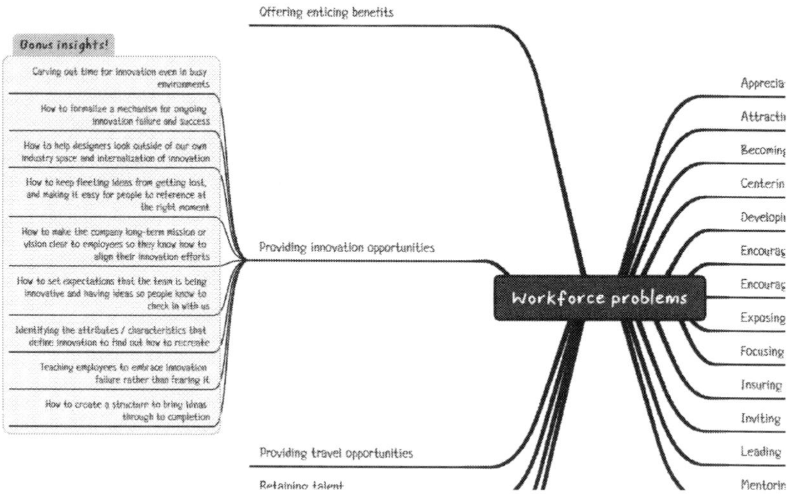

Key lesson

I'm not sure I grasped the most important takeaway consciously at the time. Only once I reflected on which themes triggered the most problems did it reveal itself.

The true lesson from this first brainstorm? If you want to engage your team with your innovation efforts, rally them around what's important to *them*.

Bonus insight

A surprisingly meta gem resulted from the nine specific problems the group identified for the theme of "providing innovation opportunities." Concerns that represented a prescient insight into the challenge we'd just embarked upon.

These will likely reflect the concerns your team will have with any fledgling innovation program you try to introduce as well. Be prepared to address them.

1. Carving out time for innovation even in busy environments
2. Creating a structure to bring ideas through to completion
3. Formalizing a mechanism for ongoing innovation failure and success
4. Helping designers look outside our own industry space and internalize innovation
5. Keeping fleeting ideas from getting lost, and making it easy for people to reference at the right moment
6. Making the company long-term mission or vision clear to employees so they know how to align their innovation efforts
7. Setting expectations that the team is being innovative, so people know to check in with us
8. Identifying the attributes/characteristics that define innovation to figure out how to recreate it reliably
9. Teaching employees to embrace innovation failure rather than fearing it

The first—and only—game

An interesting phenomenon occurs the first time you invite your team to share product ideas. You're going to have a lot of people with a pet project they've nurtured in their head for months or even years. That perfect project that they know would change the way your organization serves customers—if they only got the chance to share it.

Embrace it. It's wonderful! Leverage it during the launch phase to get everyone excited and help your team develop the habit of working on innovation projects and owning their ideas.

But it's also not enough. I learned that quickly.

In retrospect, those first couple of sessions instilled us with a false sense of success. The ideas were flowing. Team members eagerly shared their backlogs of pet projects and invited critique from the group.

But we never quite reached our desired number of presentations in any week, so we didn't feel comfortable eliminating people before we knew the full breadth of the idea landscape. The structure was flawed, and this should have been our first red flag.

Once the ideas presentations naturally tapered off, and we did narrow the field of ideas, I had the distinct impression that elimination hurt people. Having their long-held pet product concepts deemed "not good enough" stung, and they wouldn't have any recourse to reintroduce revamped versions of them for nearly a year.

But we'd committed to seeing the game through, so rather than wait for 25 ideas only to narrow to 15—as outlined in the rules—we proceeded into our first Explore cycle several weeks early with a mere handful of ideas. Designers paired up with other designers to consider their ideas from different angles.

Truth be told, the first batch of ideas was weak. But I can't blame the team for that. The format of the program just hadn't set anyone up for success.

I knew the program was in danger when the person responsible for the strongest idea of the bunch told me she was withdrawing it from the game because she had lost interest in pursuing it. My game had failed to keep her inspired, failed to instill fresh excitement about the prospect of seeing the concept through. And we hadn't considered contingency plans for strong ideas getting abandoned.

Where we failed

Our biggest problem was that the entire game concept was a locomotive on a rigid track. Rather than engaging participation, it was breeding frustration. If people wanted to get on the train once it had passed the first stage, they were stuck waiting for the next game. Nearly a whole year. And if people wanted off the train once they were on it, either because they'd lost interest in an idea or simply gotten too busy to pursue it, the game itself suffered.

Without excitement for ideas, there would be no creativity.

It was time to adjust our approach.

CHAPTER FOUR

ITERATION 1.1—EASING UP ON THE RULES

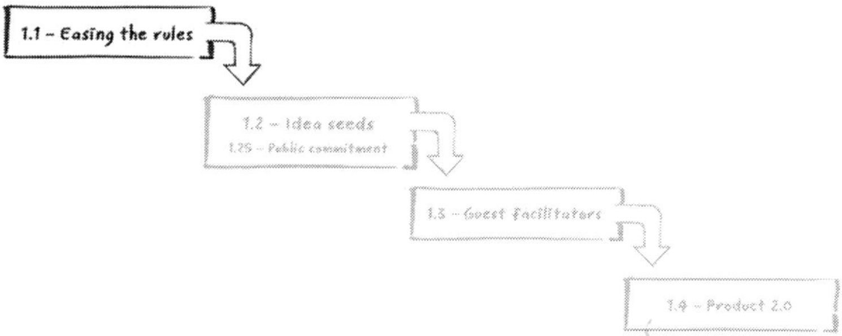

ONCE WE ELIMINATED the formality of the game structure, and with it the need for the innovation committee, my general impression was that our three team leaders almost forgot about the entire initiative. They went back to the business of managing the team, with our ongoing Creative Burst efforts simply one more thing happening in the background.

But to me, all that mattered was that no one told us to stop.

Our natural evolution kept the weekly forum for sharing and discussing ideas but dropped the forced structure of the game. We also eliminated the hassle of coordinating lunch orders by moving our biweekly meetings to 1:00 p.m. instead of noon.

Without the rigidity of the game's forced rules, once more, UX team members found themselves getting excited about ideas. They had an outlet for sharing them, and it wasn't necessary to wait until a new cycle came around any longer. So each week, designers would simply let me know in advance that they planned to introduce an idea, either via formal email or a simple heads-up in the lunch cafeteria.

Team members explored new ideas they'd started playing with, iterated ideas that had "failed" during the first game, and generally embraced a renewed sense of collaboration and creativity.

But after a few sessions without additional guidelines, the well of ideas began to dry up yet again.

CREATIVE BURST SESSION: HEARABLES

Rather than originating from a customer problem, an emerging technology trend inspired our exploration into the concept of "hearables." A portmanteau of "hearing" and "wearable," the term encompasses the universe of wearable technologies that leverage sound. Around the same time, voice-based user interfaces and assistants like Siri and Alexa were moving into the mainstream.

Goal

The trend captured the interest of one of our visual designers. He believed that if we applied our collective imagination, we could identify intriguing opportunities to solve our customers' problems using sound as both input and output.

Participants

Our rework of Creative Bursts renewed the enthusiasm for participation in the sessions, and this brought most of the team together. Of course, they also didn't mind the personal invitation from our designer.

Visuals

Our visual designer created a presentation to define and introduce the concept of hearables to the team. He shared both their historical applications—assistive hearing aids and music earbuds—and some of the emerging uses—in-ear sensors to track heartrate.

Session activity

Believing he could inspire our team about the opportunities for leveraging hearables, he delivered the Creative Burst as more of a lunch and learn. After the presentation, our designer led a discussion about how hearables, and voice interfaces, might benefit our users.

Noisy manufacturing environments where in-ear technology could trigger employees to change workstations or take their meal breaks.

Hospitals where the privacy of hearables conveying staff movements could help nurses preserve patient confidentiality.

An alternate voice-based clock-in for busy environments like retail stores, restaurants, or construction sites where multiple employees arriving simultaneously historically results in lines crowding at the clock.

Follow-up exercise

Without the formalized structure for evolving concepts from ideas to fully realized designs during the game, we didn't pursue any follow-up exercises for hearables and voice-based interactions. I assume the designer who shared the concept continued to investigate their potential—he did mention them from time to time. But as a team, our exploration ended with the Creative Burst.

Outcome

We didn't immediately embrace the concepts of hearables and voice user interfaces. But over the next year, the potential for sound-based solution design occasionally crept into our work. Most notably, they resurfaced when we reached an iteration I'll discuss later—Product 2.0.

Key lesson

Yes, we lacked a formal structure for cultivating ideas

immediately after we dissolved the game. But I think our failure to make immediate progress on sound-based interfaces stemmed from a different problem.

Technology-based ideas can't be realized as a single solution.

After all, "voice interfaces" aren't discrete functionality—they're a collection of alternate input and output methods. To explore how our team might implement any emerging technologies, we needed a framework that would remind us to ask "how might we..." while we worked on *all* solutions, innovative and mundane.

"How might we enable users to interact with sound?"

"How might we benefit from machine learning?"

"How might we leverage location awareness to automate users' tasks?"

One killer feature might showcase the greater potential of an emerging technology, but the effects will most likely ripple throughout the product suite.

CHAPTER FIVE
ITERATION 1.2—IDEA SEEDS

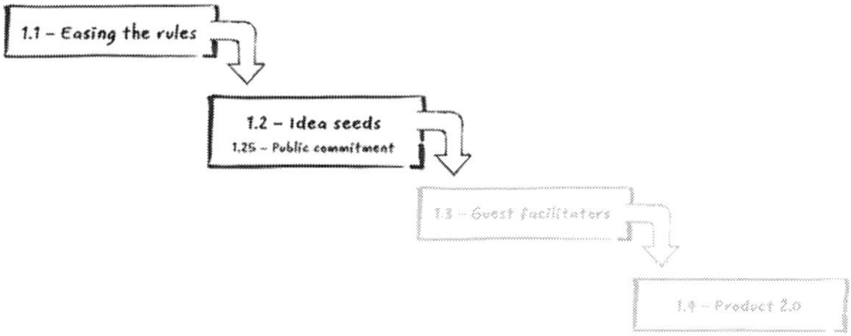

AFTER SEVERAL CONSECUTIVE sessions where I came to the Creative Burst meetings with idea starters for team members to discuss, debate, and sketch, we shifted to idea seeds as the default approach.

Sometimes, we'd work on an idea I'd dug up from our archived research findings or something I'd observed personally during a user research session. We'd review the unmet need, discuss our experiences with its impact on our existing projects, and start to ideate solutions we might implement independently or within our headlines.

Sometimes, UX leadership would share an interesting problem that they'd heard discussed outside the team and suggest that we feed it into the collective Creative Burst brain.

Sometimes, a team member would volunteer to share an compelling emerging technology, and we'd rally our insights around the opportunities for incorporating the technology into our solutions.

Sometimes—like in the following Creative Burst session about Date Selectors—I challenged us to think differently about a basic component that I thought might benefit from an original approach. We'd look at either a suite-specific common component from our library or a basic design component where our choice of technology had us lagging behind best practices.

The variety of approaches kept discussions lively. And because I spent a couple hours each month researching idea seeds, we prevented the wellspring of ideas from running dry.

ITERATION 1.25—PUBLIC COMMITMENT

Finally, we had a fresh infusion of ideas to play with each Creative Burst. But we still needed to solve the problem that ideas didn't survive beyond the session. So, despite proactive creative thinking in considering innovation solutions, we still lost sight of our primary goal highlighting the design team's ability to lead innovative product strategy within Kronos.

We had to encourage everyone back into the hard work of

innovation. We needed just a little more structure we'd lost when we'd abandoned the original strict rules along the road.

What got us back into the game turned out to be quite simple: own it.

After each brainstorming session, the team discussed the long-term potential, if any, for evolving the idea into a pitch-worthy solution. If at least 2–3 people were interested in pursuing the idea, they agreed during the Creative Burst to follow up as a team. One person took ownership and coordinated subsequent design efforts, and we added those topics to our weekly project tracker for greater accountability.

This additional accountability did the trick. Teams made ongoing progress designing solutions for ideas they'd agreed to champion. And because we now tracked projects weekly, teams had an obligation to formally justify why they had stopped pursuing the idea.

CREATIVE BURST SESSION: DATE SELECTORS

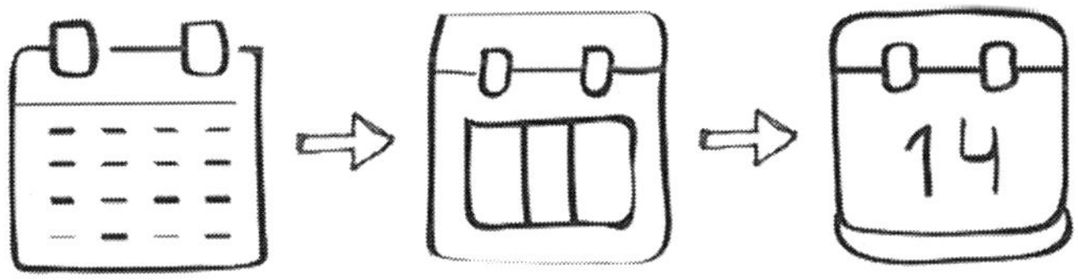

One of our most memorable sessions revolved around the simplest of idea seeds: creative ways to let users select dates and times. And I truly mean memorable, albeit not necessarily effective.

Goal

In workforce management, as in many types of software, users of all types are constantly attaching information to specific dates and times.

Managers schedule employee meal breaks. Employees request specific afternoons off to attend a child's dance recital. Human Resources introduce new gym membership benefits at the beginning of the fiscal year. Everyone receives paychecks according to a payday schedule. Executives plan grand openings for new retail locations to prepare for the holiday shopping season.

As a team, we spent ample creative energy thinking about how to better help people organize and visualize time.

So for this session, I wanted to explore one of our most fundamental interaction components: the humble date selector.

Our own date selector was an off-the-shelf component provided by our chosen JavaScript library. It operated reliably but offered only the most simplistic functionality: display the current month for the user to choose a date, or allow them to navigate forward or backward, one month at a time, to choose a different date. Fine if users want to choose a date in the current month 99% of the time, but unnecessarily cumbersome in all other scenarios.

We as a UX team had pretty much not touched it, aside from helping define the interactions for a few customizations the developers had made to support niche use cases.

I planned to issue a challenge. How could we innovate the process of communicating a chosen date to a robust system like ours?

Participants

Participation had been floundering lately, so we had a small group that day, maybe 8–10 people across both offices and those attending remotely from home.

Visuals

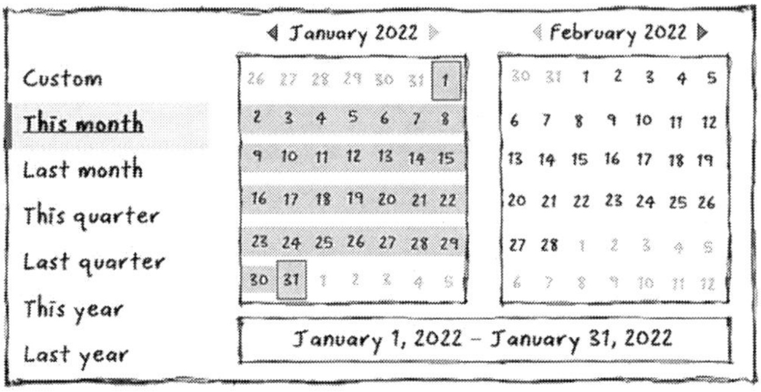

I kicked off the session showing some examples of what I thought were particularly nice date selectors from other pieces of software. Date pickers where users could zoom out to selections of years and dive back in to easily select far future or past dates. Date range pickers with semantic date ranges like "last month" or "last quarter" to ease complex selections. Date pickers with embedded data to help inform the selection process. Date range pickers optimized for mobile that supported intuitive "drag to select" interactions.

Session activity

We spent the hour alternating between segments of heads-down sketching and heads-up sharing. Which meant five minutes sketching individually in notebooks and on whiteboards, taking pictures of what we'd drawn, then I'd pull the photos up on the screen for all to see.

Each round's ideas grew more outrageous than the last. Yet the escalation of pure creativity inspired everyone to feed off each other's ideas.

Follow-up exercise

After a session of this nature, there wasn't much to do beyond catalog our photos in our archive and document every unique idea in a spreadsheet. I hadn't planned the activity with a strong "next steps" in mind. And, frankly, I was totally flustered by the direction the brainstorm had taken.

After all, time travel isn't the most...technically feasible...method for a date selector. Shooting the date "Space Invaders" style—"pew, pew, pew!"—probably isn't the most...accessible. And tattooing a day on your arm and a month on your ankle lacked...scalability.

There were simply no practical next steps for ideas like these.

Outcome

Still, the humor and overall fun of the session provided us with a much-needed revitalization. We ended up with an inventory of 105 novel methods for choosing dates. Most fell in line with the previous crazy suggestions, but several *didn't* fall under the category of "utterly bonkers."

Had I not succumbed to how thoroughly the results of the brainstorm flabbergasted me, I should have rallied a subset of the group around ideas that stemmed from novel—but still plausible—interaction methods. Yes, the potential imprecision of a slider-based date selector might initially frustrate users. But if we hadn't abandoned it before even exploring it, the unique approach might also have proven a viable starting point.

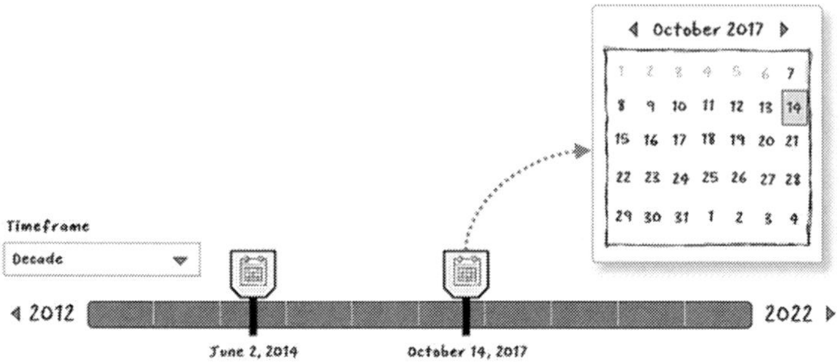

The session certainly taught me the importance of providing minimal constraints. While you rarely want to limit creativity during brainstorming, you might consider minimal guidance to keep everyone focused on designable and developable solutions.

But people talked about the Creative Burst. Fondly.

Key lesson

The occasional hour of silliness might help invigorate the creative spirit of your team, but don't let every session devolve into the ridiculous.

Oh. And dig deep for the diamonds in the rough and ridiculous. Find the gems you overlooked.

CHAPTER SIX

ITERATION 1.3—GUEST FACILITATORS

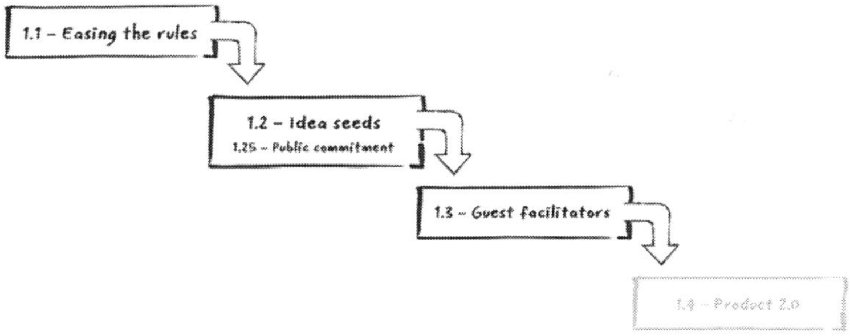

ONE WEEK, the research team requested ownership of an upcoming Creative Burst for an activity they wanted to prepare, and they asked me for help with the logistics.

The first exercise was amazingly popular.

Most importantly, it triggered team members to imagine how they might offer a more immersive exercise and present it at a future Creative Burst.

The planning and coordination to collect and deliver crafting materials to multiple offices required more effort than a normal Creative Burst. But occasional special exercises, run by guest facilitators, helped keep the format fresh and team interest high.

CREATIVE BURST SESSION: 3D PROTOTYPING

While 99% of our design work impacted screen-based software, within Kronos, the physical timeclock device held significant historical and organizational importance. At the time, both the user interface of the timeclock and the physical hardware were also undergoing extensive redesign and reimagining processes.

Taking inspiration from these efforts, the UX Research team guided us through a physical prototyping exercise to dream up additional opportunities to enhance our physical devices to make life easier when our users "clock in" for their daily jobs.

Goal

In their discussions with the folks responsible for timeclocks, the Research team had learned a lot about the challenges our timeclocks faced. In the complex modern work environment, it's not enough to simply "punch in" at the beginning of the day and "punch out" at the end. Members of our team were already tackling more robust timeclock functionality, like requesting time off and swapping shifts on the software side of things.

But some of the familiar challenges workers face when "clocking in"—like cold environments, arms full of items to restock, sanitation concerns in a kitchen—posed enormous challenges for those responsible for readying our existing hardware to meet these complex customers' needs.

Despite our lack of hardware design experience, the research team challenged us to apply our creativity to solve these types of problems.

Participants

We'd announced a "special exercise," but more effectively, the arrival of our "care package" of craft materials in the Montreal office had sparked the rumor mill. We'd successfully drummed up excitement. So that for the first time in a while 25–30 UX team members participated. The full team.

Visuals

During their introduction to the problem space, the research team shared some pictures of our hardware team's physical environment. Work desks strewn with bits of clocks—circuit boards, touchscreens, buttons, screws, and bolts.

While we couldn't work with manufactured plastic or computer chips, we had plenty of raw materials to stand in for the real thing, so we could have our hands-on experience.

Session activity

Armed with a supply of craft materials—cardboard, pipe cleaners, colored electric tape, sticky notes, permanent markers, craft paper—we spent a couple hours transforming our raw ingredients into devices our familiar personas might appreciate.

We dreamed of ways to help them balance tracking what they needed to do—clocking in and out for shifts and breaks, switching activities, requesting time off—with accomplishing what they needed to do—the daily work of manufacturing, cooking, restocking, construction.

We considered their environmental challenges. We imagined how to repurpose existing objects in their work environments—

doorknobs, tables, chairs, even lanyards—to provide functionality traditionally relegated to timeclocks in entirely new contexts.

Outcome

At the end of the session, our research team delivered photos of our device ideas to the hardware team. Dozens of photos, each an idea, big and small, illustrated the kind of insight our UX team could offer despite our historical focus on digital solutions.

Key lesson

Breaking people out of their standard modality, which for our team was screen-based design, inspires enormous enjoyment, which in turn breeds creativity.

CHAPTER SEVEN
ITERATION 1.4—PRODUCT 2.0

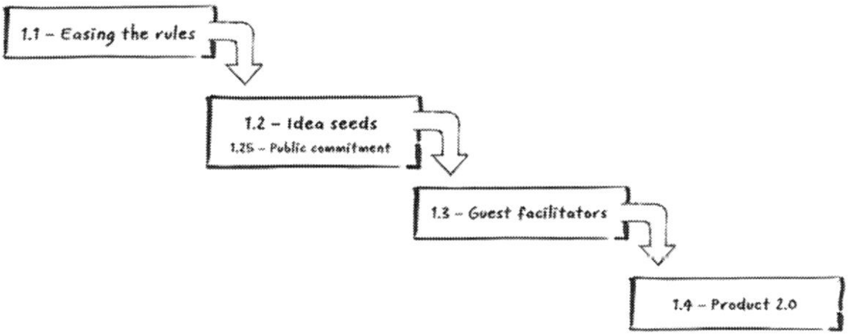

AFTER ROUGHLY A YEAR, we had internally shared Creative Burst ideas, large and small, and we were ready to package everything together and give executive leadership a sense of what our efforts had accomplished.

The initiative would be our first true foray into grassroots product strategy—the first time the UX team had proposed a holistic, design-first product vision.

Design and development were well underway for the initial release of what would become Workforce Dimensions. So we set our sights on goals for the 2.0 version of the suite. With our vision in place, and the entire UX team behind it, we hoped to set the standard for the suite's evolution from version 1.0 to version 2.0.

PRODUCT 2.0 RFP

For those UX designers who have worked with Requests for Proposal (RFPs) in a design or development agency, you'll know the rigor and formality of the process for many companies. This was not that. This was an invitation. A rallying cry.

UX leadership announced that they were seeking input from all team members on the sorts of topics, technologies, and standards we should be considering for the nebulous future of "Product 2.0."

They challenged us to share designs that would answer questions like:

- What customer and user needs had we been overlooking?
- Where had budgetary restrictions cannibalized our first-release designs?
- What emerging technologies might enhance our suite's capabilities?
- What marketplace trends threatened to pass us by?

With several weeks to prepare and a year's worth of Creative Burst brainstorming and design work in our collective back pockets, our team set to work dreaming up nothing less than the future of enterprise workforce management software.

PRODUCT 2.0 PRESENTATIONS

The tangible nature of envisioning a specific thing—the 2.0 version of our existing software suite—invigorated the team. It honed our focus in ways the free-form, ambiguous nature of many Creative Burst sessions never could.

During our Product 2.0 pitch session, we spent an afternoon sharing ideas. Each designer determined their own approach, but each proposal generally included a pitch deck and a design, shared in an 8-minute presentation followed by a 2-minute Q & A.

Pitch decks included a combination of philosophical and ideological considerations—best practices we should implement, design principles we should adopt, industry and technological opportunities we were overlooking—and framing for the design—relevant personas and their needs, research evidence, and an introduction to the design itself.

Designs ranged in execution method: sketches, low-fidelity wireframes, high-fidelity designs, clickable prototypes, physical demonstrations of solutions that weren't screen-based like hearables and location-based concepts.

Some ideas reflected polished versions of previous Creative Burst design work, while others reflected new concepts designers had crafted independently. By the end of the session, individual team members had shared more than a dozen strategic design proposals. That was the good news. The bad news: more than a dozen disparate design proposals.

Before we could pitch anything to executive leadership, we needed a cohesive strategy and message.

PRODUCT 2.0 STRATEGY

Following the Product 2.0 presentations, our three UX leadership team members reviewed all the pitch decks and designs. The goal: evaluate what they'd seen and package it—along with their own insights—into a strategic vision the team could use to guide the next phases of innovative design work.

Over the next couple months, UX leadership:

- Noted the common personas and design goals associated with those personas
- Analyzed the presentations to understand and document underlying themes
- Distilled common themes across the collection of designs
- Mapped their six common themes to our three larger organizational goals to illustrate the relevance of our work

At the end of this effort, they presented their findings and strategy to the team. Alongside each strategic theme, they displayed those key screens from team members' presentations that best illustrated it.

Despite the repackaging, every team member could see the direct path our own work had taken into the larger strategy. Nearly all design proposals lived on as components of the larger, blended vision.

To evolve these ideas into a more complete set of designs, UX leadership assigned team members from all three disciplines (the ones whose earlier work contributed to each strategic theme) and tasked us with collaboratively evolving those ideas into a more complete set of designs.

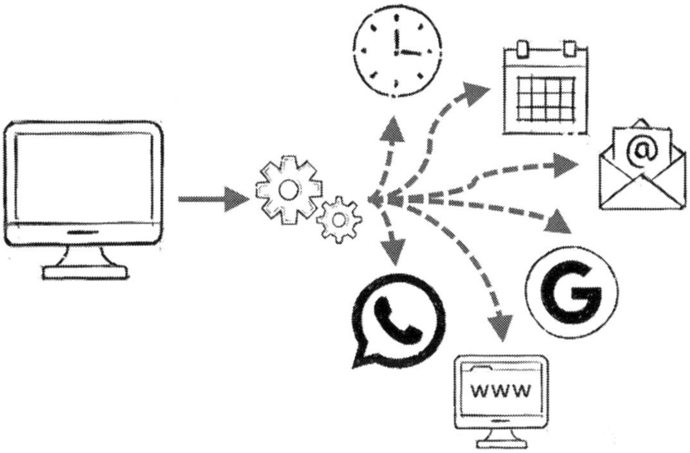

CREATIVE BURST SESSION: CONNECTORS

Facilitating Creative Bursts dominated my focus, so the program had been running more than a year when I finally managed to delve into the long-held concept I'd pledged to investigate since my early days at Kronos. The "when I could"-concept.

As I'd studied both our product suite and the challenges of workforce management, I learned the importance of managing employees' work schedules. This reminded me of my days working hourly jobs—restaurant, retail, and temporary technical work—and what I'd want from my work schedule in the era of Google Calendar. I wondered: why could our employees not embed their Kronos-managed schedule inside their personal calendars?

Following the invitation to envision the 2.0 version of our flagship product, I considered how the new application programming interface (API)—a technical capability we'd improved drastically from earlier versions of our product—might offer users control over their own information.

During the product 2.0 presentations, I shared my vision for a feature set I termed Workforce Connectors. Mobile-first, with a heavy focus on widgets, Workforce Connectors would be the IFTTT or Zapier of workforce management.

Work schedules alongside family calendars. Morning alarms automatically adjusted when traffic delays threatened to make employees late for work. Notifications sharing available shifts at a nearby store when an employee's personal calendar confirmed a free evening.

Goal

As part of executing the holistic Product 2.0 strategy, we now needed to transform my high-level Workforce Connectors concepts into a viable feature.

The biggest challenge: make a complex and highly technical concept like an API accessible to our largely non-technical employee persona.

Participants

A few of my teammates were fascinated by the IFTTT analogy. 4–5 of us gathered for a mini–Creative Burst to explore our next steps.

Visuals

Everyone had seen my Product 2.0 presentation about Workforce Connectors, but we started off the session with an informal review to level set our discussion.

Session activity

My Product 2.0 presentation shared a vision of a result—the world made possible by a feature whose nuances I hadn't yet considered. Now we needed to construct the underlying interface that would ultimately enable users to mash together workforce data with their own.

Since we hadn't prepared inspiration images in advance, we spent most of the session reviewing the actual interfaces of our favorite IFTTT-like services—automation and workflow services—to identify a baseline experience: IFTTT, Zapier, Automate.io, the Android-based Tasker and its Apple counterpart Workflow.

But most services like IFTTT, which itself is an acronym for "If This Then That," assume users have at least a minimum understanding of technical and programming concepts like conditionals. And IFTTT is even the most consumer-friendly of all of them. Most, like Zapier, target users with an even stronger business and technical focus.

Our solution needed to not only elevate the usability of the interfaces far beyond what existing services offered but also enhance the accessibility of the underlying concepts. No easy task.

Follow-up exercise

We wanted to embrace the challenge of creating a better workflow and automation tool, but we couldn't do so in the abstract. First, we needed to identify plausible, designable content. Following our first session, we each set out to define a complete workflow—everything from the initial trigger to the ultimate goal, and every step in between, mapped to the source of the relevant information.

So I imagined and documented my user story.

My goal? Arrive to work on time. What does "on time" mean? 15 minutes before the start of my shift. My workforce data contains both the time my shift starts and which store I'm working at today. Current traffic plus my current location determine how long my commute will take. My personalized settings tell the system how long

I need to "get ready for work." My smart alarm combines all that information and determines when I need to wake up.

Using story details like these, and other similar scenarios from my teammates, we started to see the bigger picture of how third-party information, workforce information, and personalized settings would come together.

We just needed to back up a step and design the interface to make it all possible.

Outcome

I truly wish I could say we crushed this challenge and designed something brilliant—the potential for game-changing user value was immense. In truth, this was one of the projects that simply fell by the wayside after three or four collaboration sessions. As each of us, myself included, focused our attention on a single project following the Project 2.0 strategy, this one simply did not make the cut.

Key lesson

Like the promising game idea whose owner simply lost interest, my lack of stick-to-it-iveness for my own pet idea spoke volumes. Unfortunately, I haven't yet determined what those volumes truly had to say. I think it's most likely that we needed a stronger project leader to advocate and excite us to continue work on a casual project. In retrospect, I might have tried inviting a trusted product manager to help champion our cause. His fresh perspective might just have provided us with a much-needed infusion of enthusiasm.

CHAPTER EIGHT
CREATIVE BURST CAPSTONE EFFORT "TASK COMPLETION"

"TASK COMPLETION" was one of the projects that lived on as an even deeper strategic initiative after we no longer held formal Creative Bursts. Starting with this project, everything came together: the core principles embedded in our innovation practice from the start, combined with the lessons that forced us to evolve along the way.

The topic is one that UX practitioners in many fields will find relatable: how to enable users to track progress towards successful completion of their task.

Our legacy product had a "wizards" feature, which only a handful of our customers used. It was complex. It required the support of our implementations team to configure. But the few customers who used wizards relied heavily on them.

For a while, we knew the project to redesign our wizards would be coming my way—since before we even began our innovation initiative. But we had yet to receive the official green light to move it into our regular project queue.

To prepare, our research team had completed a customer survey and a series of contextual walkthroughs with promising survey respondents earlier in the year. For our exploration into task completion, this proved a valuable foundation.

Despite the pervasive problem, the concept wasn't officially part of the current work stream (and thus had no tight deadlines). Good reasons for my boss to suggest introducing the concept to the team within the framework of a Creative Burst to see if we could do something special with it.

So I laid the groundwork across several Creative Burst sessions for our most strategic design effort yet.

CREATIVE BURST SESSION: TASK COMPLETION PROBLEM INTRODUCTION

Everything in a modern work environment revolves around the completion of tasks. Whether you work in a white-collar profession, as I'm sure many of you reading this book do, or as an hourly employee in a restaurant, or as a gig economy worker offering people an alternative to owning a car in a busy city, we spend our professional existence completing some task that someone has assigned to us.

Sometimes, tasks create the very scaffolding for our jobs, like the steps in the software development process. Sometimes, tasks supplement our jobs, like remembering to snap a photo of the food we just delivered. I think we all feel the burden of the need to manage, remember, complete, and report progress on those tasks.

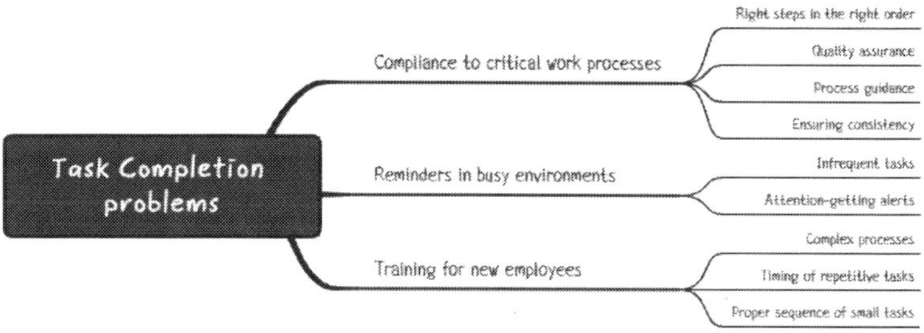

Goal

Task completion would be a relatable challenge for our team. After all, at times we all feel the weight of pending tasks bearing down on us.

Now we needed to take a step back to break down the problem space and understand our users' task-related questions and concerns.

How do I know what I've done so far? How do I know when to take the next steps? How will I know when my tasks are complete? How will I report task completion to someone who needs to know (my direct manager, perhaps)?

Can I get my tasks to tie into my external goal-tracking systems, like a more specialized to-do list? What external information—like personal scheduling commitments—should I consider to fully understand availability for my assigned tasks?

How do I understand progress across all of my tasks?

And the most complex question of all: how do I jump between single-step to-do items, multi-step task flows, and collective tasks that require performing the same smaller task repeatedly?

My goal of this introductory session was to give the team a sense of the inherent complexity of our customers' task environments and inspire people to reimagine how our software addressed these issues based on this activated empathy.

Participants

Probably three quarters of the team showed up for the initial presentation—maybe 12–15 designers and researchers. I had teased the problem space, and people were intrigued.

Visuals

After re-analyzing our earlier research results, the research team and I identified three broad categories of problems our users relied on wizards to solve. Despite the small sample size, the breakdown felt comprehensive.

- **Training**—Teach new employees how to do certain tasks until they were comfortable proceeding without support of the wizard.
- **Compliance**—Ensure employees complete critical tasks according to company policies.
- **Reminder**—Help employees remember to complete certain tasks on certain days.

I prepared slides introducing these three categories, plus the primary personas impacted by managing tasks, and a general overview of what we knew about the problem space.

Additionally, one final slide acknowledged the possibility that we might not know enough. But it was ok. We had a foundation of information, and we could draw upon a wealth of individual experiences. We'd accept our assumptions, proceed to identify use cases, explore design solutions, and validate later.

These slides eventually formed the basis of our pitch deck.

Session activity

Due to the complexity, presenting the problem space and inviting discussion took up the entire session. Although only a few team members had been exposed to our legacy "wizards," most did have anecdotes to share about the kinds of tasks users faced using other products they'd worked on.

We all knew our users had a broader fundamental need for task guidance, and we had a tremendous opportunity to go beyond what we could accomplish with a simple redesign of a legacy feature. To offer something innovative and amazing.

Follow-up exercise

To preempt the team from getting stuck in the narrow solution space suggested by our legacy wizards, I challenged them to find inspiration from the outside world. I told them not just to reflect on their own task management solutions, but also to broadly consider more general issues of tracking progress and getting reminders:

- What applications do an excellent job helping you visualize your progress?
- What novel interactions alleviate the burden of tasks?
- How do your favorite tools nudge you to complete tasks at the right moment?

Outcome

Mission accomplished. We had inspired the team, and they dug into the problem space and truly reimagined how users could understand and complete the many complex tasks and work obligations our system created.

Key lesson

Giving people permission to charge forward in the face of so many assumptions seemed liberating. The palpable excitement stood as an important reminder that even if you don't know as much as you might like, you have to start somewhere.

CREATIVE BURST SESSION: TASK COMPLETION INSPIRATION

In the days leading up to the inspiration sharing session, team members often stopped by my desk to let me know they'd saved screenshots or short videos onto our share drive. I hadn't anticipated the sheer number of visuals or the team's enthusiasm.

Goal

Yes, the fundamental goal of sharing inspiration was to further inspire. However, the deeper goal of seeking excellence outside our own product space was to elevate design thinking and avoid the common patterns and ruts all of us fall into so easily when we're overwhelmed by a challenge. We reminded everyone that we ultimately wanted to achieve the same superior design we admired in our everyday apps. At the project's outset, we set a powerful precedent.

Participants

I was pleasantly surprised at how strong the retention rate was following the initial problem presentation. Of our original 12–15 people, almost all returned, eager to present the inspirations they'd collected.

Visuals

This session needed no more visuals than the ones contributed by team members. Whenever possible, people shared screenshots from their favorite applications on their favorite devices highlighting the best examples of task completion, progress tracking, reminders, and alerts. When a static image wouldn't convey the delight of a transition or interaction, people had captured quick videos to share with the group.

Session activity

In turn, each participant shared not just their inspiring images. Why did they love it? How had the app or website won them over?

Here are some anecdotes:

One team member shared the Domino's pizza tracker, inspired by the real-time transparency of progress updates and the playful—albeit completely optional—interactions with the illustrated environment.

Another team member shared Duolingo, blown away not only by how seamless the learning progression was but how gently it reminds users to keep at their efforts. Not to mention how fantastic the illustrations conveyed visual branding and kept the learning environment lively and engaging.

Someone from the Montreal office shared lesser-known options for planning ahead for public transit in Google Maps, and how clear the visualizations made choosing the best route at any given time of day. Since most of us based in the Massachusetts office drove everywhere, it was our first time looking at the public transit options.

I shared a screenshot from the end of the Backerkit project survey, inspired by how they handle the moment in the process when you've completed your task. Yes, it's a celebratory moment of accomplishment and a community-building opportunity for backers to cheer on project creators. But Backerkit strikes a balance with the task-oriented awareness that backers might still need a reminder of the details they entered, along with easy access to reopen the survey and adjust their responses.

We saw examples of great onboarding, guided processes, straightforward task management apps and more quirky ones like

Habitica, and apps to help people follow productivity methodologies like Pomodoro®.

Follow-up exercise

For the group, the only follow-up exercise was a challenge to revisit the inspiration images and start to consider how they might translate to designs within our own problem space.

My task was to make sense of the trends I'd seen across all inspirations and distill useful features and approaches we might leverage.

Trends like:

- Gamification and playfulness as engagement strategies
- Visualizations of multi-step processes
- Success screens that instilled a sense of accomplishment commensurate with the effort of the task
- Just-in-time reminders that don't aggravate

Outcome

The exercise gave everyone an opportunity to share not just things that inspired us, but also a little insight into our own mindset and that of our colleagues.

The library of inspiration also acted as a great frame of reference, providing a shorthand for our own ideas as we worked through later design processes.

Hindsight adaptation

Although we kept the sharing simple, this would have been a perfect opportunity to create an inspiration board in a creative tool like Prezi to share and later review our screenshots.

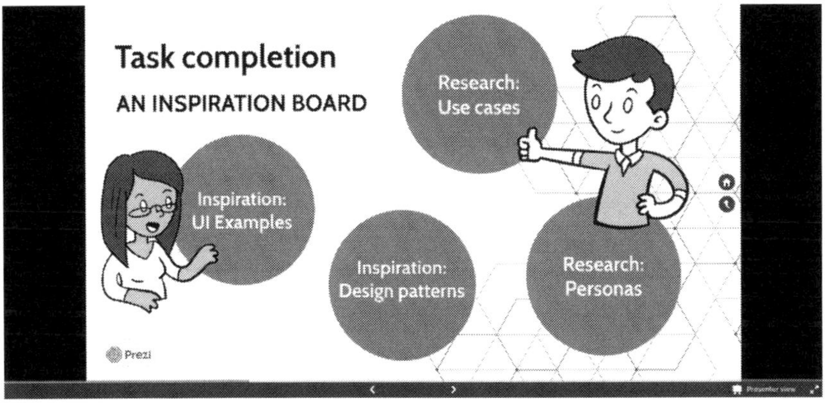

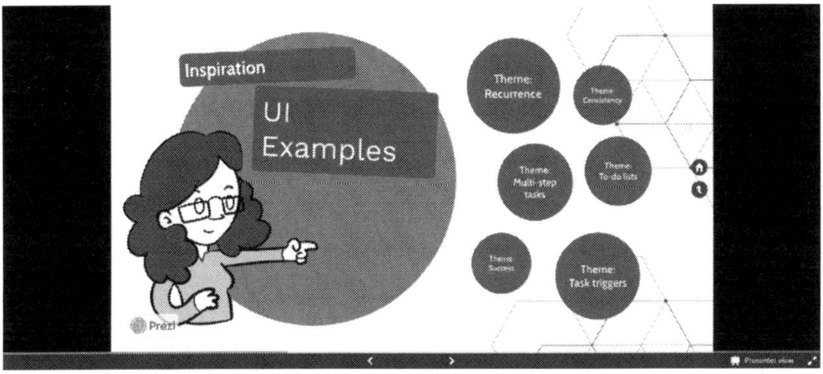

Key lesson

Inviting everyone to share the applications they use every day really connected with people in a way I hadn't anticipated. Gaining exposure to brilliant design and interaction solutions was an asset by itself; just as effective, if not more so, was the inspiration from seeing what mattered to each other.

CREATIVE BURST SESSION: TASK COMPLETION SKETCHING

With the stage set for creativity, I rallied the UX team around the process of designing a solution that we could see through to completion.

Goal

In addition to the obvious goal of kicking off the design process, this project held a deeper goal for me personally.

Our first six months of Creative Bursts began with a failed game structure before proceeding through a series of interesting-but-ultimately-dead-end design challenge brainstorms. People were enjoying themselves, and I saw evidence of them bringing a newfound sense of strategic empowerment into their day-to-day work. Some of the ideas generated during Creative Bursts even began seeping into other projects.

But freshly into the new year, I needed to prove we could *finish* something.

Participants

The first brainstorming session included maybe a dozen team members, still brimming with passion for dreaming up ways to help our users, guide them through some of the more complex workflows in our software.

We wouldn't keep our researchers once we moved into deeper design, but they were always good sports about participating in preliminary sketching sessions. I think they enjoyed dusting off parts of their UX training they'd left behind when they chose to focus on user research.

Visuals

Eager to begin the brainstorming process, I figured beginning the meeting with slides would only bog it down. We breezed through the inspiration images one more time as a refresher, then I kept a summary of the trends I had identified on the screen as "idea joggers" throughout the session.

Session activity

The session was as straightforward as a sketching brainstorm gets.

We sketched. We shared. We sketched some more. We shared some more.

Throughout the hour, people tied their concepts back to the original inspiration. They remembered more inspirations and iterated their designs. They mashed up concepts and broke through obvious solutions. They explored different facets of the problem and sketched distinct phases of the process.

By the end of the hour, we had generated more than a dozen unique sketch concepts, and I had five team members eager to pursue the problem through its next phase of solution design.

Follow-up exercise

With the core team of five designers, and another half dozen people ready to review progress and offer critique at key milestones, we now had to transform our initial sketch concepts into a few detailed screens that illustrated the central task flow.

We each mashed up the results from the design brainstorming session into cohesive concepts and shared them in a meeting several weeks later. With original inspirations still shining through, we referred to each concept by a clever name like TurboTask or Workforce Pizza Tracker.

Outcome

After several mashups and iterations of these core workflows, we ended up with one design solution we felt captured the best of everything. Approximately four months after I first shared the problem with the entire team, I pitched the solution to our newly promoted VP of UX.

The concept of task completion didn't make it onto an immediate roadmap. Yet a few months later, it profoundly inspired one of the central principles of a product visioning exercise. From there, the concept would morph into the foundation for one of the most strategic features we'd tackle during my time at Kronos—a smart form of task completion support we called Assistants.

Key lesson

Sure, you can't make an omelet without breaking some eggs, but you also eventually have to cook them! The only way to achieve your goals is pushing through the discomfort and uncertainty of trying something completely new.

CHAPTER NINE
OUTGROWING CREATIVE BURSTS

FROM OUR PRODUCT 2.0 visioning exercise, we came away with a clear product strategy, marching orders from our UX leadership to explore and realize that strategy, and a reinvigorated sense of design ownership. And with our Creative Bursts reaching the "it's been over a year" mark, it was time to, once again, have a closer look at our innovation program and reevaluate our approach.

FEEDBACK SESSION

What roughly marked the one-year anniversary of our program coincided with moving to a new office. Learning our way around a new space constituted enough internal disruption. We held off on any major process changes for a few months, including evaluating the Creative Burst, until we'd settled in.

When we finally did decide to hold a guided session to gauge how the team felt about the program, we received several pieces of feedback that would ultimately change the course of our practice.

Some team members had concerns about whether we were truly innovating rather than simply realizing feature improvements. Most loved the hands-on sessions and wanted more of them. One person challenged whether we were better off handling innovation like an incubator with a smaller dedicated team.

But the biggest single concern—described several ways, from loss of momentum to wanting a tighter cycle between idea, design, and sharing—was the common feeling that half a day every other week simply wasn't enough.

While we were still *enjoying* the biweekly collaborative brainstorming and design sessions, feedback suggested they were insufficient, and our Product 2.0 efforts had proven that we no longer *needed* them. The team had successfully internalized the underlying principles, so without constant reminders they continued to "think innovation."

Paired with the increased visibility of our efforts—and our stake in the ground about future product strategy—we now primarily needed to execute. We needed something tangible to demonstrate we could provide more than just preliminary ideas and rough sketches. To prove when our ideas introduced hard problems, we could also think them through.

INNOVATION PROJECT ALLOCATION

So after nearly a year and a half, we eliminated the formal structure of biweekly group sessions and the single afternoon of work time every two weeks.

What replaced it?

Ongoing innovation project allocation.

Every designer, interaction and visual alike, would have the freedom to either pursue a single innovative idea of their own or participate in a teammate's project as part of their typical 3–5 project workload. The research team would support our efforts, planning and scheduling research for those innovative design projects—like

Assistants, the evolution of Task Completion—alongside our standard roadmap-driven headlines.

PRODUCT 2.0 EXECUTION

Freshly motivated by the design strategy we'd help develop, most team members chose to collaborate on projects inspired by the Product 2.0 themes, guided by our UX leaders.

Unlike the roadmap-driven projects from product owners that served an immediate need and supported our first major release, these Product 2.0 projects we generated ourselves were hallmarked by ambition. Budget wasn't a concern yet. Nor was feasibility of execution.

We aimed to

- explore an idea fully;
- execute at the highest level;
- be relentless in evaluating if a design would truly make our users' lives better;
- offer more to the overlooked audiences our software already impacted;
- make smart assumptions, yet form plans to validate;
- stretch the technical capabilities of our solutions;
- push past the failure and keep at ideas we knew had promise;
- develop concepts we could pitch.

Without the safety nets of requirements, road maps, and project timelines, we had the freedom to do those things. But during earlier iterations, this freedom had also been a curse, and it often left our team members spinning their wheels in early or middle stages of design.

We marched forward, enabled by just enough constraints to keep us productive: focus on a single UX-led project at a time, weekly accountability tracking, and strong project leadership.

INNOVATIVE PROJECT PARTNERSHIPS

However, the most encouraging and exciting result of our efforts over that first year? More teams understood our team's strategic design capabilities.

Not only did executive leadership acknowledge what we'd accomplished, but other teams envisioning innovative solutions began contacting us to partner with them.

The grassroots innovation spirit was taking hold company-wide, and people now realized: if they needed design support for their ideas, big or small, we were the team to help.

Product management started coming to us with pet projects they'd had on their wish list for years, eager to have the visuals they needed to convey the concept to partners on our customer panels.

UX-minded developers recognized opportunities to improve where they'd been forced to compromise on features and started to ask our input during functionality reviews.

And data scientists who found interesting trends in customer data asked us for visualizations that would bring that data to life.

The goal? To determine if customers would be willing to pay for a product that surfaced those trends—as actionable information.

WHAT INNOVATION EFFORTS STAYED?

While the Creative Bursts formed the core of our innovation program's first phase, their primary purpose was to keep strategic design top-of-mind. To provide a structure for encouraging team members to do more, think deeper, and stretch our definition of what it meant to deliver design at Kronos.

As designers integrated principles into their day-to-day thinking, other initiatives naturally emerged, and those lived on even when the Creative Bursts gave way to what was next.

Here are just a few examples:

Design Day Out

UX leadership implemented a quarterly "design day out" where our entire team gathered in an off-site collaboration space. While each event had its own goals and agenda, in any given session we might review and critique recent work, audit for inconsistencies and resolve competing patterns, discuss exciting emerging technologies, share project and external learnings, plan for future initiatives, or propose and debate team process changes.

It was an opportunity to step away from our standard offices and daily obligations, connect as a team, experience a new creative environment, and ensure we stayed in touch with each other's work. The objective was to gather everyone in the same room and reconnect, but we did occasionally need to include individuals remotely. The special focus of the day was more important than the physical location.

The Safe & The Crazy initiative

Without using the name, I hinted at "the safe & the crazy" initiative as one of the strong signals that our team was ready to dive into this journey. While this initiative coincided with the origin of the Creative Burst program, it emerged independently.

Targeted at Interaction Design team members, the initiative stood as a reminder for us to always push boundaries. On every project, we reminded ourselves to deliver not just the frictionless design solution that met the stated requirements but at least one more "crazy" solution.

"The crazy" might be a solution that looked at a problem sideways—and thus solved the same underlying need—but didn't necessarily conform to an expected approach. It might be a solution that incorporated an emerging technology our software didn't currently support.

No matter what the new approach might be, part of the challenge was to treat both solutions equally when presenting. We could certainly acknowledge the difference between "the safe" and "the crazy," but we would always justify how our design decisions tied to user goals for any solution we presented. We didn't ever include "crazy" solutions purely for creativity's own sake.

Peer 2 Peer, Pairs on Projects

Reminiscent of the practice of pair coding, our UX team fundamentally shifted from assigning a single interaction designer to each project to having at least two senior team members, depending on the size and prominence of the headline. Large projects might include another 1–2 junior team members. Strategic or high-visibility projects even included as many as three senior team members.

Larger teams challenged assumptions by drawing on different perspectives and background experiences. We reduced design pattern conflicts that previously emerged from solving the same problem over again in different ways, because each project benefited from visibility into more team members' external projects.

Multiple team members participating on projects was, however, the largest innovative result. This aspect elevated the level of work by reducing the likelihood that a single person might settle for their first idea.

Team members pushing each other to look beyond the obvious, and not just for a single "crazy" option, resulted in better work, plain and simple.

Open design exploration

Once we gained freedom of schedule and experience with the

diverge-and-converge approach to brainstorming, we unlocked the opportunity to go deeper. Yes, managers still assigned a core team to projects—usually 2–3 interaction designers, 1–2 visual designers, and a researcher— but for particularly interesting problems, the core team was authorized to invite anyone from the team to participate in the early design stages.

The core team would prepare a presentation describing

- the problem they'd been tasked to solve;
- the central use cases;
- any other relevant details that might entice others to participate.

Then they'd plan a series of design explorations, which included activities like inviting people to share inspiring concepts and patterns from external products, or generative sketching to push past obvious solutions into the truly unique. These larger group sessions comprised the divergent design phase.[1]

The core team would take the dozens of results from earlier sessions and converge on a few outstanding concepts, which they'd then share back with the larger group. After a second round of divergent thinking, where the larger group of participants generated additional design variants from the new starting points, the core team would then move into their usual design process with more unique ideas than they could have generated on their own.

THE FUTURE OF INNOVATION AT KRONOS

Full disclosure, I moved on from Kronos during the summer of 2018, therefore I truly don't know how—or, technically, if—the innovation program has continued to evolve. Based on anecdotes about team structure changes after I left, I choose to believe that the UX team doubled down on strategic design and innovation.

But that's just a guess—what has actually taken place on the UX team, especially in light of the 2020 merger with Ultimate Software and the formation of UKG—is a story for another person to tell.

However, one of the company-wide initiatives even called for a "Culture of Innovation" by the winter annual meeting in January of 2018, around 18 months into our design team's Creative Burst experiment.

And at the time I left, teams actively pursued several strategic innovation projects to solve a variety of customer challenges not then covered by the core product suite. Designers infused the principles from Creative Bursts into day-to-day projects and continued to advocate for deeper customer insights. Other teams actively sought collaborations with UX to help realize ideas they'd come up with.

I trust they continued to ride the momentum of our early successes and embrace the opportunities we'd proven they could handle. In short, UX-driven innovation had a promising future.

PART THREE

CREATIVE BURST TAKEAWAYS & GENERAL PRINCIPLES

CHAPTER TEN
INNOVATION PROGRAM FOUNDATIONS

> Hindsight is notably cleverer than foresight.
>
> —*Chester W. Nimitz*

LOOKING BACK on our Creative Bursts, we identified three building blocks that are critical when an innovation program's success means creating value within an organization. If your program incorporates these building blocks, you'll be well-positioned to achieve the organizational impact you want for your team.

I'll discuss this in the subsequent chapters.

But these building blocks aren't the only critical elements. Without these other logistical ingredients—I'll call them "foundations"—the program will fall apart. They have nothing to do with delivering innovative ideas but have everything to do with running the process smoothly.

Here are some of the key elements we needed; depending on how you structure your program, I expect you might need them as well.

FOUNDATION 1: TIME & SPACE

Your team might not have executive buy-in if you're looking to *prove* the value you can provide. But deep thinking and design work take time. To spend that time, your team needs the freedom to step away from other projects—and this requires a minimum buy-in of permission from your UX leadership.

You need to spend a few hours here and there pursuing a design concept. You need to grab a few teammates in a conference room to review nuggets of wisdom from a recent research study. You need to share emerging trends and dream up ways to incorporate them into your solutions.

If you're working through lunch, you need food.

FOUNDATION 2: PROGRAM LEADERSHIP

You need a passionate individual committed to running things. Team interests wax and wane over time, in alignment with the team members' own natural cycles of motivation. You will need your program leader's enthusiasm and preparedness to persist and spark new enthusiasm from the team.

FOUNDATION 3: FACILITATION

The program leader will be the primary facilitator. But for your program to exist beyond one person's tenure as leader, or even to continue while the leader is on vacation, the team needs a solid foundation of facilitation skills. A team needs to understand different phases of design brainstorming and encourage divergent or convergent thinking as the situation demands. A team also needs the experience, and tact, to balance all voices and be comfortable enticing insights out of team members who don't traditionally speak up.

FOUNDATION 4: LOGISTICS

We certainly learned the hard way how much work our internal event coordinators did when we started running an informal program at lunchtime without their assistance. We designers usually scheduled our own meetings, so we were familiar with booking conference rooms for small groups. But we rarely had to coordinate large conference rooms and never had to order lunch delivery for a group of 25 people.

During my leadership of the Creative Bursts, I found myself handling basic logistics like

- ordering lunch;
- booking large rooms;
- sending participant invitations and reminders;
- coordinating guest speakers and facilitators;
- connecting remote participants & requesting help from IT when issues arose.

Basically, I was responsible for handling—or coordinating the people who would handle—anything that it takes to run large-scale meetings on a regular basis. By ensuring that the process ran smoothly, I freed the team from annoyances like fighting with teleconferencing technology and gave them room to focus attention on sharing ideas.

FOUNDATION 5: DESIGN COLLABORATION TOOLS

Every company has their pet tools, UX teams sometimes even more so. From design tools like Adobe Creative Suite, Sketch, and Figma, to internal team wikis on Confluence or SharePoint, to project management in Basecamp or Asana, to team discussions in Slack, to email and meeting scheduling in Outlook, we spend our professional lives jumping from one tool to the next.

You'll need these general categories to support your day-to-day logistics. These are in addition to the design tools you use to develop your ideas into shareable solutions, your suite of UX research tools, and your internal communication tools like email, teleconferencing, and chat.

Whatever your team's favorite tools may be, they should be accessible to your entire team. Ideally, they should even be accessible to other teams, once you're ready for that kind of visibility. Bonus points if they're searchable and allow metadata like categories and tagging.

Those tools include:

- library of team-shared inspiration and education
- storage for problems
- repository of photos and assets from group sessions and design collaborations
- sharing platform for design solutions

Optional, but potentially helpful tools:

- online whiteboards
- mind mapping/flowcharts for early brainstorms and organizing thoughts
- visual feedback platforms

Beware: tools research can also be an unfortunate trap. If you're worried your team might be prone to finding "the perfect tool" as a diversion, start out assuming less is more. Focus on building the habits and instilling the foundational principles of grassroots innovation in your team.

Start with basic solutions that leverage your team's existing tools, like designating a folder on a shared drive for program assets like photos or spreadsheets of problems. Address upgrades to more robust tools only when your team keenly feels their absence, especially if your company has a rigorous or convoluted software procurement process.

Even my later example Creative Bursts assume the most fundamental tools to keep you from venturing down this particular rabbit hole, but I expect you'll substitute in your favorites, or research more purpose-built solutions over time.

CHAPTER ELEVEN
BUILDING BLOCK 1—PROBLEM IDENTIFICATION

Sometimes the questions are complicated and the answers are simple.
—Dr. Seuss

THE FOUNDATION of this entire process is a UX team's ability to identify real problems. Our program organically attracted problems to solve from a variety of sources. The benefit of hindsight allows us to determine which sources and sorts of problems were most effective.

Then we call upon every UX skill we have, regardless of formal team member roles or titles, to deconstruct and understand the problems. Only then do we try to solve them.

SKILLS

As designers, when we work on roadmap-driven projects, someone typically hands us the problem we need to solve. Although we ourselves may need to uncover nuances of the problem, or do research to understand it better, someone outside the UX team has already identified a problem, formed requirements to solve that problem, and delivered those requirements to us.

By contrast, when UX teams drive the product strategy and innovation design process, it's up to us. We identify and clarify the problem we want to explore with a designed solution.

As UX team members, we encounter problems every day in our work: sitting in research sessions, talking to customers directly, or—if we use our own products—sometimes experiencing these problems firsthand. To uncover problems intentionally, document them, dissect them, and determine which have the greatest potential to create value within our organizations—that takes some practice.

Like any of the skills we cultivate throughout our careers as UX practitioners, we simply need to better understand problem identification and develop a system to practice it. The good news is that the underlying skills we need to identify high-value problems are many of the same hard and soft skills we exercise and hone throughout our professional careers.

Skills like:

- curiosity
- empathy
- relationship-building
- qualitative analysis

By pairing those foundational skills with some curated activities, you will end up with a sustainable backlog of problems ready for you and your team to design solutions for.

ACTIVITIES

If your organization has yet to embrace formative research practices and relies solely on UX research for evaluative usability testing, you might need to get creative identifying problems that impact your users. Then again, if your organization has yet to embrace UX research at all, you're probably already used to that.

Here are just a few research practices we relied on. If clumsily handled, each poses its own problem, but in the hands of talented, senior UX practitioners, they can offer enormous insight. Most of what you uncover means you have to accept some assumptions for now that later need to be validated, but some combination of these should provide a solid starting point.

Watch the marketplace

Tread carefully with this one. Environments ripe for disruption rarely reward "just do what the other guy did." You're looking to extrapolate problems, not borrow solutions. Careful analysis of trends in the larger marketplace can reveal a need that your competition is trying to meet—possibly even failing, if you're lucky—and should provide excellent opportunities for different solutions to the same problems.

Leverage public sources of information

Information gleaned from sources analyzing your marketplace or your software—like mobile app reviews or blog posts—can sometimes be superficial at best, but you can often tease out deeper insights from analyzing the gestalt of what you find. If you work on software that historically requires extensive training, you can often find your own customers' how-to guides for their employees with a simple Google search.

Talk to your customer service/technical support teams

Every day, your users reach out to teams within your company to complain about their problems. Some of them might originate from bugs—things that simply don't work as they should and will be fine once fixed. But many questions and problems stem from issues like confusing interactions and missing features. This is a reliable source. Uncover how your internal teams document, share, and escalate the problems they learn, and see if you can gain ongoing access to their resources.

Mine your existing research

If you've ever sat in on a usability test, you know that—even for the most talented facilitator—it's impossible to keep participants 100% on task. They're never as concise as you wish, packaging insights into quotable soundbites of feedback. They are, after all, human.

But offhand comments and friendly discussion while you switch screens or pull up a new prototype not only create rapport. The resulting "extraneous" or "unrelated" is also often an insight you can mine later, even if it's not relevant for your current study.

If your research team saves findings beyond the project-specific reports, you and your team members can find unmet needs and nagging problems your users shared with you weeks, months, or years ago.

If you're storing your research findings and participant quotes in a database, you're probably already ahead of most companies. If not, you'll need to rely on more cumbersome, time-consuming methods.

Here are some ways to proactively glean customer problems from your existing research efforts.

If you're a designer:

- Sit in on research sessions and take notes.
- Watch old research session videos.
- Review research findings reports and catalog trends.

If you're a researcher:

- Invite extra note-takers and assign them to specifically document comments that don't relate to the current study.
- Gather as a team to discuss trends unrelated to specific study goals.
- Tag videos with interesting comments or seeds of ideas for future reference.

Identify trends in your customers' data

If you have a data science team, reach out to them and be proactive. In my experience, data scientists tend to be curious and persistent statistics wizards who love pushing the boundaries of what machine learning can teach them about the data. Also, they are extremely practical-minded and care deeply about what the data can teach the people who created and own it—your customers.

So if your company has a long history of customer data, and has a data science team mining that data, what are the odds they've uncovered potential insights that could benefit from your team's design thinking methods? Excellent! As a bonus, executive leadership should be predisposed to accepting a design solution inspired by insights from this strategically focused team.

Consider emerging technologies

The future is a direction, not a destination.
—Edwin Catmull

While technology-first solutions can be a potential trap, they can be an effective framing method to identify problems if you consider solutions that leverage emerging technologies.

Apply your deep knowledge of your personas: observe how your own users adopt emerging technologies—or even mature technologies that your company simply hasn't incorporated into your solutions yet. How does the technology help users solve problems? What information delivery methods have customers adopted that your company hasn't considered?

Could you provide novel solutions using technology like voice, augmented reality, virtual reality, location awareness, solar power? Do users want to connect your software to their smart home? Will they want to bring your software to Mars?

Try to be a futurist. After all, if your company succeeds in avoiding killer disruptions, you'll need to serve your users in the world of the future.

Draw from personal experience

Many of us work in environments where our UX teams design solutions for people unlike us. But as UX practitioners, we cultivate empathy for a living. Even if we're not actually the target audience for our solutions, we understand our users. We can draw parallels between our lives and some element of the daily challenges that bring them to our software.

Yes, you will need to validate solutions based on these problems. Rigorously. But don't dismiss two other valuable sources of problems to inspire innovation solutions: your experience and your instincts. After all, Henry Ford apocryphally pointed out that if he'd asked people what they wanted when he invented the automobile, they'd have told him, "A faster horse."

PITFALLS

A challenging part of exploring innovation inside a larger company is to identify the right problems to solve. Too big, and you might stray too far from your organization's goals to achieve the buy-in you'll need. Too small, and you might find yourself working on obvious features that product managers already earmarked in the roadmap for future attention. Too many assumptions, and you might end up designing solutions that don't reflect actual user needs.

Here are some of the biggest pitfalls to avoid when identifying problems, and some tips for applying due caution.

Superficial or excessive analysis

Assumptions hold a special importance in the world of a UX design team. We never know our users as well as we'd like to. We spend our careers forced to make decisions and define solutions based on incomplete information. But this is our reality. A hallmark of being a successful UX practitioner is the ability to get creative about how we fill in the gaps.

However, we must strike a balance between analysis paralysis, where we linger too long trying to learn more, and settling for a superficial understanding of the problem space in which we seek to design solutions.

Hedge against both extremes of assuming too much and getting bogged down with unhelpful learning with a few tricks.

- Include multiple designers on a project to challenge each other; determine whether they are jumping to unfounded conclusions or whether they are simply using the need to learn more as an excuse for creative resistance.
- Bigger up-front assumptions require more validation, so design in small phases and test frequently to refine your underlying principles.
- Set an aggressive timeline for your early phase designs. You can always elaborate later, but time pressure can, counterintuitively, bring out creativity.

Failure to move forward

Especially when they get good at uncovering problems, team members can find themselves stuck at this stage rather than moving into the demanding work of seeing innovative design through to completion.

It might be fine to designate a subset of your team as "problem finders." Your research team likely enjoys this part of the process and is well suited to the task.

If your UX leadership will oversee projects and mentor everyone throughout the design process, they might take ownership of problem identification, or at least problem filtering and selection.

Whatever your approach, make sure you define clear roles and set expectations for everyone at each stage. Impress upon your designers how valuable their efforts are with later phases of crafting solutions. But make sure you inspire genuine buy-in—you need creativity from your team, and lack of intrinsic motivation will block it.

Overabundance of freedom

One key takeaway from our own problem identification process was a reminder that the paradox of choice[1] is all too real. Too much freedom of "what we *can* work on" often leads to increased intimidation of the blank page.

While good UX practitioners have recognized the importance of constraints as a mechanism for honing creativity for ages, those boundaries originate in the problem identification phase. Innovation can aim to solve a large problem, but sometimes scaling back a team's focus to a smaller problem yields disproportionate results.

After all, disruptive influences start eating away a company's market share when an upstart solution reinvents how customers see one of the smaller problems your product addresses. Be the one to reframe customer problems with your own innovation, and you'll not only gain new customers who have only that problem, you'll also start insulating your organization against losing your existing customer base.

By keeping your problems bite-sized and having leadership initiate key projects when interest lags, you should continuously revitalize participation.

RESULTS

The result of your problem identification effort can certainly be a single idea for a single solution. But with some extra effort on the part of your innovation leader, you can—and should—create a repository of problems.

Define guidelines for how team members should document each problem. Provide examples that illustrate proper grammatical structure. Determine what metadata will help—things like associated products, specific personas, categories, source of the information—so members of your team can drill down and find types of problems that most excite them.

Create a shared Google Spreadsheet, build a database with a tool like Airtable, post a page on your team wiki or intranet, share a document in your team's Slack channel. Whatever your team's favorite method of sharing information, capitalize on each person's individual efforts to provide an ongoing source of inspiration for the entire team.

You might eventually want to create a scoring system for evaluating problems according to the potential for value or likelihood of achieving executive buy-in. Likely you will need to do this for a while before you get a feel for those things. Document your overall progress as your innovation practice lives and breathes. What you learn about the trends you see emerge: feed that back into the "problem identification" stage.

CHAPTER TWELVE

BUILDING BLOCK 2—SOLUTION DESIGN

THIS IS THE "COMFORT ZONE" for most UX teams, and ours was no different.

But designing solutions for fuzzy problems calls for a different approach than designing to meet a specification or a set of requirements for roadmap-driven projects.

We managed to blend the varied inputs—design inspiration, knowledge of our personas' needs, diverse team skills and life experiences, calculated assumptions, and research data—into design solutions that we believed would provide value for our customers.

SKILLS

This book doesn't aim to teach you how to be a UX practitioner creating design solutions for the first time. After all, the very premise of this book assumes that you are on a design team—or leading a design team—and eager to prove that your team can provide value beyond the current opportunities you have on a day-to-day basis.

If you and your colleagues are indeed capable of that kind of value, you're already part of a team of talented UX professionals. That's not to say you don't have junior team members or interns who are just starting on their professional journey. I merely assume that more senior team members are already mentoring them in the standard skills while also instilling a mindset of innovative design.

Therefore, I'm not providing an exhaustive review of all the skills you'll need. I assume you already know how to do things like design sketching, wireframing and prototyping with your preferred software, developing information architectures and process flows.

What I will call out are some of the more advanced or specialized design thinking skills UX practitioners will need to leverage in their innovation design practice.

Skills like:

- writing user scenarios
- storyboarding
- divergent & convergent thinking
- design critique

If your team is anything like ours, you have people well-versed in skills like this but don't get as many opportunities to exercise them on your roadmap-driven design projects as you'd like, where tight budgets and timelines often reign supreme. You often wish to supplement your everyday design skills with more of the specialized skills. When you do that on projects where you control the pace of your forward progress yourself, this is the innovation design work you and your team can build a reputation on.

ACTIVITIES

The biggest challenge for us, once we had the freedom to pursue our own projects, was overcoming the urge to proceed with "business as usual." Our team consisted mostly of senior designers who excelled at churning out quality design work—leveraging our standard component library—that met documented requirements. Of course, we urged our project teams to meet standards of quality and to always put customer needs first. But for most projects we trusted the expertise of our product management team and rarely questioned the underlying goals of the projects they handed to us.

With this new breed of UX team–driven projects, though, we needed to slow down and push our design explorations beyond the first obvious solution that sprang to mind. If you work in an agile or particularly deadline-driven organization, you might similarly find slowing down to be one of your team's first challenges.

There's no reason that you can't still work in rapid iterations and validate small pieces of functionality often. This isn't necessarily the "big design up front" that so many agile methodologies demonize.

But it does require a cognizant effort of switching between divergent and convergent design thinking rather than settling quickly on a single solution. You must deliberately avoid designing the local maximum, and instead seek the best, most innovative solution to explore before proceeding with your in-depth design process.

As you pursue that objective, here are some activities that might help you break out of your standard design process.

Create assumptive personas

Even if your organization has invested significant resources into persona development, and even if your personas already constitute an important part of your design process, you might find that your innovation design process reveals opportunities to reach new audiences. If these new audiences aren't represented by your current personas, assumptive personas can be a great tool for incorporating your new user base into your design process.

Our Creative Burst process pushed us to explore multiple audiences that we'd traditionally lumped into a single persona. So we paired our existing knowledge with some careful assumptions based on demographic and psychographic nuances to further segment one existing category into several more specific personas.

As we developed our persona hypotheses, we carefully

articulated our assumptions. So that when we eventually planned our research to test the effectiveness our designs, we would remember to include questions to validate those underlying assumptions.

Share problems

The simple act of sharing problems and discussing them as a large group can yield an abundance of hidden information. While you prepare to delve into a specific problem with your solution design process, talk about the problem with as many people as possible. The moment you choose to promote a problem from "just another on the list" to "let's try to solve this!" is the moment when you need to know as much as possible.

Go back to the source of the problem and learn more. Was it an insight from the customer service team? Who else might know more? Have you developed trainings to overcome a gap in your software? Does your sales team have to justify why the gap exists, and how do they ease customers' minds?

Armed with that kind of—and lots of—information, talk to your team as a group to uncover all the facets of the problem. Ask for input from team members who have worked on other projects serving the same audience, or even on a product that neglects to solve the problem you want to explore, learn what they know. Why haven't you solved it yet? What organizational factors have deprioritized it or overlooked it altogether?

Gather inspiration

If your design solution ultimately wins executive approval, and thus its own formal budget and roadmap, that is the moment to worry about tightly adhering to corporate branding guidelines and leveraging existing design systems. Until then, ignore those sorts of existing constraints.

Many companies revamp organization-wide creative art direction only every few years. Even companies who aren't reworking an entire enterprise product suite over the course of years, as we were, still drive their agile design processes according to existing creative guidelines. You know that by the time your product or feature launches, it will already lag behind the marketplace unless you've just undergone your most recent update.

Don't fall victim to that limitation in your UX-driven projects. Use these projects to showcase your best. At this stage, aim to push the boundaries of design. If you anticipate an upcoming trend will greatly enhance your customers' experience, or even find your existing products fall short of current best practices, err on the side of being bold.

Proactively collect examples from the outside world that addresses the problem you want to solve. You're not looking to borrow anything wholesale for your new solutions but rather define a creative benchmark.

Seek inspiration from applications that you and your team personally love, even those entirely unrelated to your problem space. What makes them so engaging? Why do you use them every day? What makes you smile?

Look for public data sources or technical capabilities with APIs that might relate to your customers' unmet needs.

Create a library for these sources of inspiration so your team can reference them as they proceed with the design process.

Ignore limitations like technical constraints you've heard from developers. Assume you'll acquire any personnel, software, or hardware infrastructure you'd need to enable new capabilities, like machine learning or voice interactions.

Set a precedent for the kind of design you want your team to be known for. Sell a vision. Compromises come later.

Map customer journeys

Oftentimes, zooming out to the bigger picture can reveal an

innovative approach to problems impacting customers. Especially if your organization produces enterprise software, you might have a particular way of dissecting customer needs into discrete products, solutions, and teams. And over time, that method often overlooks the connection points between products or the customers' larger experience.

Customer journey maps highlight different problems across a customer's entire relationship with your company and identify where your existing products serve various needs. This holistic view can uncover the biggest opportunities to translate those needs into new and innovative design solutions.

Much like any new personas, these might be assumption-based to start, but discussions within your team and with other teams, as well as during customer research, will help you refine your journey map.

These also make a powerful visual when the time comes to justify your design solution during a pitch, especially if your organization doesn't already do much journey mapping.

Create mashups

Mashing up two seemingly unrelated technologies, problems, or concepts with each other is a tried-and-true method of innovators.

What if you could track your pizza delivery the way you track your own progress through traffic on Google Maps? DoorDash and GrubHub.

What if the process of learning a language was as much fun as playing a game? DuoLingo.

To start exploring your own mashups, list out a sampling of the inspirational interactions, technologies, data sources, and design approaches you've gathered. List out various customer problems you've researched. Choose a random scenario and list out the stages of the experience, like visiting your bank. Or eating at a restaurant. Or celebrating a friend's birthday.

Pick 2–3 items from all your lists and see how the juxtaposition of disparate concepts triggers new opportunities and ideas. Once you've captured ideas from this pairing, do it again with another scenario.

Diverge first, converge second

I've mentioned diverging before converging, and related concepts like local maxima, several times. But what do I mean by it? If you're not already familiar with this in design brainstorming: it's essentially a way to avoid settling for the first, obvious solutions.

After all, the obvious is the antithesis of innovation. No matter how obvious an innovation might seem in retrospect, before someone finally created it, no one had figured out how to.

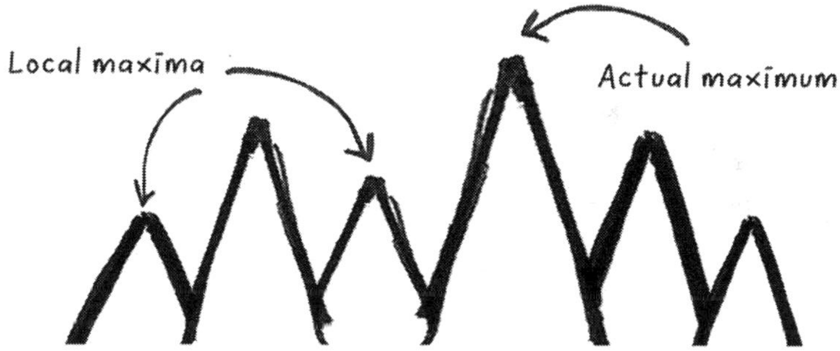

Local maximum simply refers to the highest point in a segment of a 2D or 3D geographical space. In practical terms, reaching the top of a medium-sized hill would be an example of a local maximum. In problem-solving terms, it relates to our tendency to optimize for the best possible outcome within our current field of view. The danger is in forgetting that our current view might be incomplete, and we might only be able to achieve the actual best solution if we first reframe our perspective.

When we brainstorm concepts or sketch designs to solve a problem, the first few are always obvious. In our designs for roadmap-driven work, especially if we're building on a long history of understanding our customers, these instinctive solutions often get the job done.

But for innovation design, we need to purge the obvious from our brains. A purge to free ourselves to discover truly novel solutions. Whether you subscribe to generative design, the "Five Sketches" methodology[1], or simply gather your UX team into a large room and instruct them to sketch rapidly for 10 minutes, the key is to encourage people to push past the first solutions, past the obvious.

Diverging design is all about coming up with a bulk of rough design solutions. Multiple designs per participant, multiple participants. Once everyone has at minimum 3–5 designs—even just thumbnail sketches—everyone shares their work.

Don't discuss individual designs as you share them. Simply take it all in. Note similarities and trends. At the end of the sharing, discuss what you saw and converge on a few promising solutions, even if they combine elements from multiple early concepts. Aim for 3–4.

Once a core team or a single designer has expressed the narrowed exploration with a new batch of sketches, share with the large group again. Invite everyone to create variations of these 3–4 solutions, combined with both inspirations from the first round and any other ideas they've had in the meantime, and generate a brand-new batch of 3–5 designs each.

The exact numbers don't matter, as long as you focus on pushing past the obvious. Generate lots of designs before you narrow your focus to a few promising ones, then diverge again before converging on the solution you eventually pursue. You can certainly complete a third cycle of diverging and converging if you're feeling ambitious, but make sure you do it at least twice.

Hold design critiques

Designing solutions might be the comfort zone for most UX teams, yet if you're reading this book, you must want more for your team than the usual products of that comfort zone.

Beyond methods for generating original solution concepts at a project's outset, teams must progress through the entire design process with innovation in mind.

Design critiques can help drive teams to think deeper, push harder, consider more creative solutions. Whether you choose to elect a committee of innovation mentors or keep the entire team involved, you'll need a periodic review process to keep designers from falling into their usual comfortable patterns.

When giving critiques, team members certainly can assess the quality of the design work relative to a recognized benchmark, especially if you're finalizing visuals for a pitch. But the real goal is to validate the in-progress solution as an innovation project.

Establish a set of questions to ask designers. You might ask:

- What is our inspiration?
- Have we compromised any of our original vision? If so, why?
- Does this truly create value and solve a customer's problem? Or are we just doing something different for the sake of being different?
- Have we validated any portion of the solution with customers? If not, what's our plan to do so?
- Are we limiting our perspective too much and are we ignoring problems customers see as inextricably linked to this one?

- Are we tackling too many things at once rather than focusing on the core issue?
- Is this too similar to a solution we already offer?
- What will compel leadership to buy into this solution?

PITFALLS

Designing solutions is your team's raison d'être—reason for being. Therefore, your biggest pitfalls won't stem from your ability to execute. Instead, watch out for failures to execute at a high enough level or failures to achieve what your leadership might consider innovation.

Settling for the first solution

I've already addressed this issue at length in previous sections, but it warrants repeating, since it is one of the primary pitfalls of any innovation design process. Without proper rigor during the brainstorming phase, your UX team–driven solutions might fail to achieve the quality and creativity you need to prove your team's potential to initiate strategic projects.

Getting stuck at the superficial "idea" stage

Much like we did in the early days of our own Creative Bursts, you might find your team stalling out in the superficial idea phase. If you see this happening, talk to your team.

Chances are they're either confused about how to move beyond that phase of the process, or they're feeling unsupported in some other way, like missing collaboration time, or external pressure forcing "real" projects to take priority.

If collaboration time is the issue, mitigate by having your innovation leader formally guide the process and facilitate activities like the ones outlined in this book. If you hear reports of inconvenient prioritization, ask yourself the hard question: do people truly have the time and permission they need to work on innovation projects?

Fear of failure

Good judgement comes from experience. Experience comes from bad judgement.

—Jared Spool

One of the few execution-focused pitfalls you might encounter stems from a fundamental fear of "getting it wrong." Yes, the goal might be daunting—prove our strategic capabilities to executive leadership—and the expectations for quality might be high.

Remind your team: innovation demands the cycle of trial and error—failure and learning, assumption then validation.

In the early phases of solution design, reward bravery, not quality. Remind team members that quality and design excellence will be a byproduct at the end. Early on, you need to work through the details of an idea, and that process is messy.

Above all, instill a sense of safety in your team: all critique is in service of making the team, the ideas, and the solutions better.

RESULTS

The results of your solution design phase could be a set of wireframes, a prototype, a lightweight coded solution (if you've involved developers), high-fidelity designs, a comprehensive storyboard, a video demo, or some other deliverable you've deemed appropriate.

You might have already tested the solution with users, or you might be seeking executive buy-in before you do so, depending on your organization's policy of talking to customers.

You might have shared your solutions with other internal teams and iterated based on feedback already.

But while your own personal "definition of done" for innovation projects will vary based on what you believe your leadership team will respond best to, the result of the design solution stage should be a complete set of whatever visuals you need to support your pitch.

CHAPTER THIRTEEN
BUILDING BLOCK 3—OPPORTUNITY VISIBILITY

Luck is what happens when preparation meets opportunity.
—Seneca

THIS IS where the magic happens.

Formal structure of product leadership usually exists outside UX teams, yet our team was eager to break out of the role of "followers in the process" that plagued UX's fledgling years. To steer the strategy and innovation for our robust enterprise suite of workforce management software, we followed the only roadmap we knew—trial and error.

Rather than lamenting over UX's secondary role in providing product strategy and vision, we owned the responsibility for sharing the vision we created. We used a few different methods for sharing and saw varied degrees of success.

But again, hindsight revealed insights into the key role that packaging of ideas and storytelling play in increasing the likelihood that ideas would transform into product strategy. We also identified essential skills and methods that helped ensure that decision makers would take notice to give projects their best chance of getting funded. After all, a great solution to a user's problem can't help anyone if it never gets built.

Not everything we tried worked. But one technique that proved successful was aligning our UX-driven proposals with the industry trends already on the minds of our executive leadership team.

Therefore, your opportunity visibility stage includes three major components:

1. Identify topics your executive leaders are currently interested in.
2. Hone your presentation skills.
3. Pitch, pitch, and pitch again.

You are certainly welcome to pitch all ideas once they've evolved enough to share. After all, it's impossible to anticipate when you might reach an executive with the right idea at the right moment because of his favorite personal cause or something her teenage daughter just mentioned was important to her class. If you choose to let the readiness of the solution drive its own visibility, you can skip my first component.

But executives are busy people. If we want to increase the likelihood of them accepting future meetings with us when we have something new to share, we, in turn, need to take the steps to maximize the relevance of our solutions. I highly recommend doing the extra research and being extremely selective of which ideas you pitch—at least until your program starts to gain recognition and traction within your organization.

Yes, you might end up storing completed design solutions on a shelf as they wait for their golden moment. If so, use this time to increase the appeal of your solution when you do finally present it. Validate with customers. Share with internal audiences for preliminary feedback. Refine the designs further. This especially is why many of the activities I share don't directly result in the ultimate goal: pitching to executives.

Nevertheless, get ready. Those solutions you find that already align with topics that interest your executive leadership? Get ready to sell design like you've never sold design before. Unless, perhaps, you've founded a startup and pitched for venture capital. Because that's essentially the task ahead of you—requesting startup capital for your design solution from those who control the company coffers.

SKILLS

Pitching what amounts to a startup within your organization requires an enormous amount of skill. Not everyone is comfortable public speaking, and many people don't even want to be put into positions where they need to give presentations.

As UX designers and researchers, our careers force us to cultivate a minimum comfort level to present our research findings or share the rationale of our designs to product teams and colleagues. However, some might feel that minimum comfort level is enough.

Not everyone wants to be a thought leader speaking at conferences. Not everyone wants to deliver the difficult pitches. As someone leading an innovation practice, discuss with your team how everyone wants to handle this.

Do you choose to maintain a core team of people with key skills to handle the pitches for the entire UX team? Do you challenge everyone to push through the fear and handle pitches for their own projects, even if they're still learning? Do you balance both options by making sure that each project team includes one of the more experienced presenters—to either deliver the pitch or to mentor the project owner?

These are decisions you'll need to make with your team based on their individual goals and comfort zones.

But no matter who you decide will own the responsibility, delivering these types of presentations requires skills like:

- design storytelling
- executive presentation design
- public speaking
- salesmanship

Even if you choose to leave the final pitches in the hands of a core team, everyone on the team can benefit from activities that help craft the narrative and refine the content for your audience.

ACTIVITIES

This isn't a book on creating great presentations, although I do recommend reading some of those as part of your learning process.

The activities for this building block are more nuanced because of the complexity of all that goes into creating brilliant design pitches carefully crafted to speak to your audience. You're almost going through your building blocks again as you prepare for your pitch, before culminating your efforts in the actual presentation to your executive leadership.

Whether you choose the indirect method, like sharing with other teams informally during a lunch-and-learn session, or pitch directly to executives, your team needs well-honed presentation skills. Key activities inherent within the building block "opportunity visibility" all improve some aspect of crafting those presentations—either tailoring the message to the audience or delivering the pitch.

ACTIVITIES: DETERMINE WHAT'S IMPORTANT TO YOUR LEADERSHIP

Identifying your users' problems helps you design solutions likely to excite your customers. Similarly, if you can determine what trends are important to your organization's leadership, this will help you select innovation solutions likely to excite them enough to earmark funds in the budget for further design and development.

Read your organization's and your leaders' blog and social accounts

Large organizations that maintain a blog or active social media presence often adhere to an editorial calendar. This calendar reflects topics intended to set the organizations apart as thought leaders within the industry they serve. Keeping an eye on your company's blog should therefore help you identify the trends, whether marketplace or technological, that your executive leadership considers important.

And because strong leaders of top organizations are often passionate about their industry, their individual professional social presence can similarly provide a wealth of information. If one of your design solutions relates to a recent blog post or series of social media posts, be sure to mention it during the pitch.

Review internal executive presentations

Does your company give periodic internal presentations on topics like quarterly earnings, current growth areas, and industry trends? If you attend such meetings remotely, record them so you can review for insights you can reference later. Or reach out to speakers after the presentation with a relevant comment and request the slides they shared.

Create a modified prioritization matrix

A typical prioritization matrix breaks a problem space into four quadrants, with two axes of measurement. One axis represents effort, from low to high; the second axis represents value, from low to high. Software organizations often map new products or product features along this matrix to determine where to invest development budget. Features with low effort and high value are "no brainers," while features with high effort and low value are "probably not worth it."

Features with high value that also require high effort need careful consideration.

In this case, you're looking for design solutions that have both high **customer value** and high **organization value**, so you modify the prioritization matrix with those as your axes. Low effort ideas would be a bonus. But here you're trying to make a case for why the effort will be "worth it," so your ultimate goal is to bring to life that which is truly innovative. Use your other research into executive leaders' interests when determining each idea's company value.

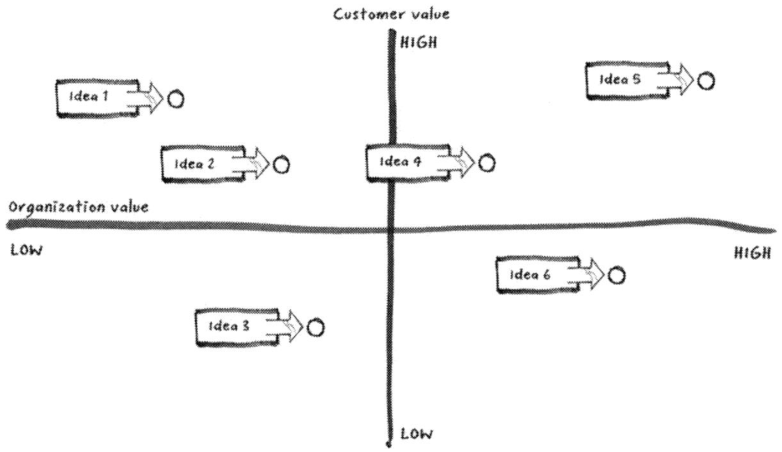

You won't share your prioritization matrix in your pitch—it's just an internal decision-making tool when deciding which designs to pitch first.

ACTIVITIES: LEARN TO CREATE AND DELIVER GREAT PRESENTATIONS

Even if members of the team choose not to deliver pitch presentations directly, participating in activities like practice presentations can give everyone solid critiquing skills to help others prepare for the big meetings.

Watch TED Talks

TED Talks aren't just a great source of education and motivation, the short format makes them a stellar case study in delivering high-impact information in a tiny package. Watch them as a group to kick off a Creative Burst session or a design brainstorm, encourage team members to share favorites in your repository of inspiration, challenge people to analyze what makes certain ones so effective. The more exposure everyone gets to great talks, the more people will start to absorb the principles.

Experiment with design presentation methods

If you have a standard format you use to present designs for typical roadmap-driven work, start experimenting with different options. If you always present low-fidelity wireframes, try capturing a video of you using a clickable prototype next time. Share a paper prototype made from various sketches. If you typically walk your audience through screens sequentially, start your presentation with a storyboard that shows transitions.

Use your day-to-day work as an opportunity to try different methods for presenting designs, and catalog reactions to figure out what works. This will take a little longer, especially if your team uses an extensive library of templates to promote efficiency. But the effort should help your team break out of "business as usual" thinking, even in everyday work, and prepare everyone for the less frequent task of innovation pitches.

Create presentations for all projects

If your team typically presents designs by sharing template-based design documentation or simply walks through the wireframe or prototype documents, spend a little time crafting a minimal presentation for every design review. You don't need to go overboard with dozens of unnecessary slides or slides for every screen you want to share. But get everyone in the habit of evaluating and articulating the core message of the presentation and crafting a few slides to share that message.

Form your own "Toastmasters" group

To get public speaking practice in general, we formed our own informal UX team "Toastmasters" group to share interesting topics in a safe environment. Any UX team member could present, and people mostly used these as opportunities to share takeaways from articles or books they'd read or workshops they'd attended. Presenters would create a few questions and solicit feedback from the audience about key areas they were trying to improve, like delivery pacing or eye contact, but also to assess how relevant people found the topic.

Consider the marketing

Many dismiss all marketing and advertising as "bad" or "annoying," yet we can learn a lot by keeping an eye out for great examples of marketing and to analyze why it's effective. Even in the initial stages of solution design, consider the taglines your organization's marketing might use to sell the solution to potential customers. What headlines might media outlets write when the product releases? Practice writing them, share them with your team, refine and rework them together.

Have a team member with background in marketing or simply a knack for marketing copy? Encourage that person to mentor others and offer feedback. If the design solution would push your organization into a new market segment, consider the name of the product. Once you've honed the marketing message and product name, use them in your pitch. Chances are these won't live beyond the pitch—your actual marketing department would take over—but it shows you've considered how to get customers excited about the product you've proposed.

ACTIVITIES: SHARE YOUR IDEAS WIDELY

Increasing awareness of your UX team's innovation design capabilities can certainly come from high-visibility victories like getting an executive to buy into adding a new product to the company roadmap.

However, you can also share ideas more narrowly and let awareness of your capabilities diffuse into the company over time.

Share ideas you're not ready to pitch. Share ideas that got shot down. You'll never know what ideas might evolve into something else with the right partnerships.

Speak at conferences

Caution: every organization has different rules about sharing intellectual property—please don't violate your company's policies, I don't want to get into trouble for suggesting it.

Just be up front about the status of the design as speculative or concept, like an automobile manufacturer sharing a concept car, especially if you choose to present at industry conferences where getting customers excited about non-existent products might backfire.

But if your company encourages you and your colleagues to share team-specific project case studies to establish yourselves as thought leaders within your industry, external attention for your design solutions might just be the inroads you need to get your organization's leadership excited about them.

Write whitepapers

When we innovate, we often find ourselves inventing solutions people don't know they need. Yet still, the solution addresses some fundamental human need that's simply being solved other ways. Tread as carefully as you would with the conference approach. But intellectual property–sharing policies permitting, writing a whitepaper forces you to articulate the value offered by your new solution, and how it would fit into the marketplace.

Deliver lunch-and-learns

Create lunch-and-learns around your designs and invite teams that would benefit from them. Did you explore a solution to a complaint customer service hears on a daily basis? Invite them to see what you've done. Did you come up with a cool visualization of customer information that would need machine learning? Share with the data science team. Have you envisioned an onboarding experience that would eliminate weeks of effort producing customer training videos and documents? Share with your client education team.

Do this often, and you won't just get other teams excited about specific solutions, you'll develop a reputation for thinking creatively about the problems your colleagues spend their days struggling to solve on behalf of your customers.

Pitch your solution to leadership

I chose to simply list what many people might consider the be-all, end-all result of this process as just another activity to highlight an important lesson—getting innovative products built is a secondary consideration in this process. The ultimate goal is visibility into your team's capabilities, and you will see that happen naturally as the culmination of the activities that comprise your innovation practice.

When you do get your opportunities to pitch an innovative design solution you've developed, keep these tips in mind:

1. Craft a story around your solution that explains who the audience is, what problem they need to solve, how your solution addresses their need, and why your company should supply the solution.
2. Once you've honed the story, build your slides. Keep them simple but provide enough detail that the slides can tell your story for you once you've left the room.
3. Practice presenting. Practice again. Make sure transitions between slides and designs go smoothly, especially if you don't simply embed screenshots in the slides.
4. Tell executive leadership what action you want them to take. If you want funding to build the specific solution, then say so. If you have another objective, like budget to hire another UX researcher devoted to formative studies to uncover more problems like this, say so.

The rest you'll figure out, as we did, by trial and error. Learn what techniques are most effective for your executive leadership and share your experiences—positive and negative—with the rest of the team.

PITFALLS

Let me first assure you that "flubbing a pitch" is not a pitfall of this process. Expect it to happen. At some point, it might. Having an idea shot down also is not a pitfall. That will, I anticipate, be the most common outcome.

But failure is a crucial step in the innovation cycle. Failure teaches us what to do better next time. Prepare more, craft our message more carefully, align our solution better to executive interests.

Whether or not your design solution ever moves beyond this pitch into a more formal experimentation phase, or even into a standard software development cycle, depends on a lot of factors out of your control. Hence the reason this is the last phase I discuss.

Your company culture might need to evolve further before your team gets more than a polite pat on the head and "that was interesting, great work!" Without a formal incubation mechanism, executives might just not know what to do with your idea. Or your solution might just stray too far from the company's current business goals to make sense for investment. There's also a huge element of luck to all of this, especially if you're proposing truly innovative or "out there" solutions.

The true pitfalls for your innovation practice are those that prove toxic to its long-term success.

Getting discouraged

The innovation leader, along with UX leadership, bears the responsibility for keeping the team motivated and positive, even in the face of repeated failures. First, accept that you can't avoid team frustration entirely—it's a natural reaction when someone rejects a passion project you've poured your precious time and creative energy into.

Once you've accepted that, do your best to minimize its impact on your team's enthusiasm by celebrating every victory, no matter how small. Have conversations to air those frustrations as a group, where you can keep it from spiraling, rather than degrade the creative atmosphere by allowing whispered exasperation.

Remind people of the importance of failure in the innovation process. You might end up discovering that "no" just meant "not yet."

Do what you can to minimize how much of the work ultimately goes to waste. Create an internal repository of design solutions—a sort of team portfolio—on the company knowledge-sharing platform. Or perhaps you find that you can use the kernel of an idea, or an interesting new component from it, in your conventional, roadmap-driven projects.

Duplicating efforts

When you belong to a team of bright, engaged designers, in a company full of similarly bright, engaged employees, where everyone focuses on serving the same customer base, you might learn that the solution you just pitched matches an idea another team has been trying to gain traction on.

While you could interpret this information as a reprimand for duplicating efforts, and thus "wasting time," you should instead interpret this as an opportunity to partner with folks outside the UX team. But even getting told "someone is already working on this" doesn't have to be the end of your idea's journey.

Perhaps their existing solution lacks refinement because they didn't have the benefit of experienced design thinkers. Maybe their solution explores a different element of the problem space than yours does and partnering will result in a more robust solution. Perhaps their underlying solution is actually better because they understand the problem more thoroughly, but they need your help crafting visual elements and coordinating research to validate with users.

This pitfall is hard to avoid entirely in large companies, even those actively working to improve communication and break down silos, because everyone is just so busy. It's impossible to hear everything that everyone is working on, especially if you're physically remote from other teams.

As you increase the visibility of your effort, other teams will naturally start to loop you in, and you'll experience fewer instances of "someone is already working on that." Expedite the process. Reinforce the bridges to other teams that you're building during your problem identification phase—proactively. Reach out on a regular basis to ask what people are working on. Follow up with early previews of design solutions inspired by a problem you learned from them. Introduce external teams to the UX team portfolio you create so they can check in occasionally.

RESULTS

The ultimate goal of your innovation practice is to increase visibility. Leadership needs to see your UX team's capacity to design high-value solutions that either serve your organization's existing customers or bring in new, tangential customer groups. Which means that your entire process has been building toward activities that will accomplish that increased visibility—presentations.

Pitch quality solutions to prominent individuals within your organization, and you will achieve recognition. Persist in seeking—and subsequently earning—that recognition repeatedly, and others in the company will start partnering with the design team earlier in the product visioning process and inviting you to participate in strategic planning.

And even if every pitch you make gets rejected by executive leadership: everything your team members invest in improving their design storytelling and salesmanship will benefit their long-term careers.

CHAPTER FOURTEEN
DIY CREATIVE BURSTS

AT THIS POINT, you've absorbed a fair number of my own Creative Burst stories and the results we achieved, and seen a framework for running Creative Bursts that might kick off an innovation program of your own.

I've discussed my own Creative Burst experiences in broad strokes so you can see how we iterated and eventually outgrew our first structured format. I've drilled into the details of our takeaways with a series of general principles and associated key activities so that you have the building blocks for a program of your own.

I'd like to share a few example Creative Burst sessions, so you can see how the process ties together. This example follows a single design project across three Creative Burst sessions, each drilling into complementary early-phase activities in the design process.

But first, how do you know what activities constitute a Creative Burst vs. some other type of creative meeting?

IS MY MEETING A CREATIVE BURST?

For the first time in this book, I'm going to give you the most infuriating answer a UX practitioner can give: it depends.

And leave it at that.

Just kidding.

At its essence, a Creative Burst is a meeting.

How you define the scope and agenda of that meeting depends on what your goals are for Creative Bursts, how many people you involve, and how much freedom, time, and space your team has for additional collaboration.

During my own program, I wanted to open up every Creative Burst to the entire team—or even people outside the team, if they chose to join—to keep innovation top-of-mind for everyone.

That meant focusing the hour-long sessions on activities that either kicked off a new project or held its own as a stand-alone activity. Sessions to inspire, brainstorm, or drill into problems.

Those activities alone, though, don't result in finished work. We still needed to regroup as a team, to work independently in our design software, to solicit feedback from stakeholders, and just do the kind of deep thinking that produces great design outside the Creative Burst.

If you're strapped for time, hold sessions more frequently, have a small enough team to keep everyone involved in every project, only work on a single project at a time, or are willing to curate who attends your Creative Bursts, you might choose to facilitate your entire process within the hour-long session framework.

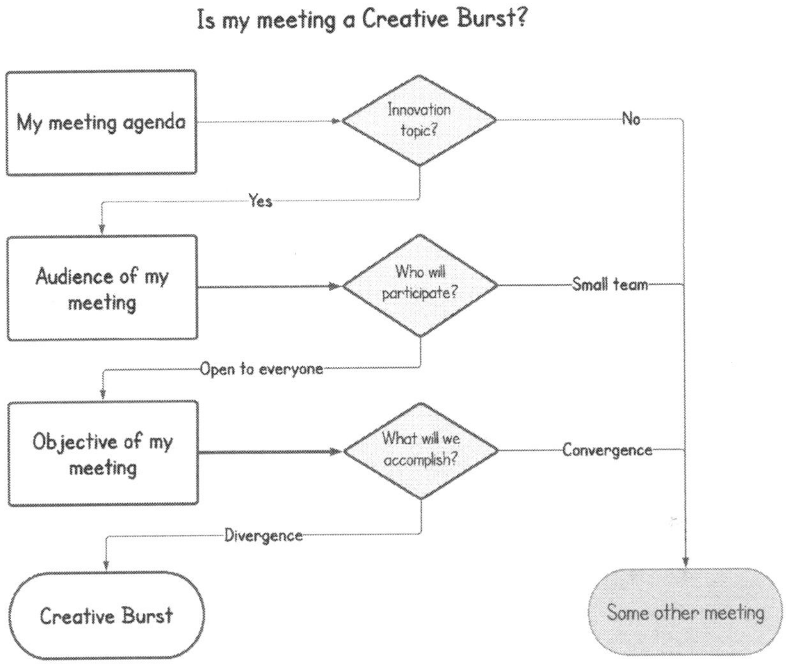

PLANNING YOUR CREATIVE BURST

I'm sure you've guessed most of the steps needed to plan a Creative Burst session throughout the course of this book, but I want to give you an easy reference.

A complete innovation program results from stringing together a variety of activities, some of which occur in the form of a Creative Burst (where the bulk of the innovative collaboration happens), some of which occur as individual and small group activities (where the bulk of the hard work gets done).

But you'll need thoughtful planning before Creative Bursts, and honest evaluation after, to maximize the effectiveness of both.

For each session, consider your

- goal;

- visuals;
- activity;
- follow-up exercise;
- outcome;
- hindsight adaptation;
- key takeaways.

Up-front planning

Your leader needs to answer the following questions while planning each Creative Burst session. If you invite guest facilitators to run sessions, work with them to make sure you have everything you need.

Goal

What is the one primary goal you want the team to accomplish with this Creative Burst?

Think big picture. This is the business or team goal your Creative Burst should achieve, not the outcome of the individual project.

Visuals

What visuals do you need to prepare?

Consider what will introduce the team to the problem space, guide their understanding, inspire them as they participate. Think about what format—slides, inspiration boards, personas, mind maps—will best convey your message.

Activity

What activity will yield your desired outcome and ultimately accomplish your goal?

Chances are that your most common Creative Burst activity will be some kind of divergent group brainstorm—that's how you maximize the benefit of bringing together diverse creative talent and focusing their attention on a single question or problem. But to constantly reinvigorate the interest of your team, vary the details. Consider my examples—things like capturing problems, sharing inspiration, sketching solutions—but also lean on a combination of activities familiar to your team and those that stretch their creative brains. Mold the activities outlined in the three building blocks into team efforts.

Then determine what materials you'll need, what telecommunications need to be in place, what real-time collaboration tools you'll use to facilitate the session. Do you need to collect and distribute craft materials for a 3D prototyping exercise? Plan ahead!

Follow-up exercise

How can a follow-up exercise bridge your efforts to either the next Creative Burst or the next long-term step in your project?

Your leader might perform a follow-up exercise to transform the results of the session into a format everyone can reference later. The group might have "homework"—like finding inspiration on a particular topic—to prepare for the next session.

Go into your session knowing what people will do next.

Outcome

What outcome do you need the Creative Burst session to achieve to move your current project forward?

The outcome—or rather, the desired outcome—represents your goal for the activity and the follow-up exercise. While your outcome should contribute to the larger goal, it might just be one step along the path.

Retrospective

Following each session, consider how it went. Your innovation program lives and breathes, and needs to evolve to continue flourishing, even through adversity. Run a small retrospective after every session, either individually or with a small group.

Hindsight adaptation

Based on the results of the session, how would you change the exact same exercise if you were to run it again?

For a couple of my sessions, I shared what I'd do differently if I were to run them today. After each session, ask yourself—and your participants—what should change for next time.

Key takeaways

What key lessons did you learn from your Creative Burst session?

Document what you learn from each session you run. The takeaways might feed back into the program as adaptations, insights into your teammates and what they respond to, general principles of innovation, etc. Even if you don't know how to apply what you learn immediately, review all your takeaways periodically and decide what they mean for the future.

EVOLVING YOUR OWN PROGRAM

Each subsequent iteration of our Creative Bursts maintained the best features of the previous one while introducing a new layer of engagement and opportunity.

When you evolve your own system, always evaluate your previous activities carefully before making wholesale changes.

You don't want to eliminate fundamental components if your team has responded well to them just to implement a necessary change. Much like we craft UX research reports to not only share the users' challenges but their triumphs as well, so that our organizations don't fix what isn't working at the expense of features users love.

Because Creative Burst is also a philosophy. A philosophy of listening to your customers and yourself, of exploring concepts in depth, of taking action on ideas, of persisting through adversity and dwindling motivation, and of seizing opportunities that arise. A philosophy of innovation, experimentation, collaboration.

Your needs will dictate the structure of your program, as will the feedback you should gather frequently from your team.

You might choose to hold isolated Creative Burst meetings as examples of how your team can infuse design thinking into everyday work. You might choose to string together recurring Creative Burst meetings into a holistic innovation practice.

Either way, you'll need the flexibility to adapt to the environment —and the team—you have.

But I can only show you the door. You're the one who has to walk through it.

—Morpheus, The Matrix

EXAMPLE CREATIVE BURST: PROBLEM SHARING

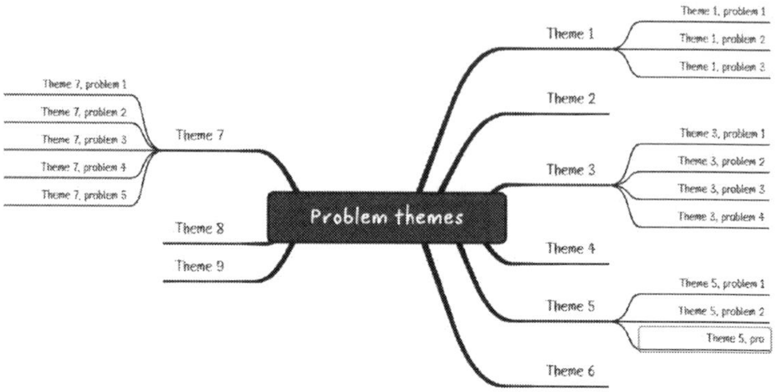

This example assumes you already have identified problems and selected one of them prior to these sessions with the larger group, or you could share two or three in the first session and let the group choose which to explore further.

Goal

Your primary goal is to relay the unmet customer or user need that you've chosen to resolve with your team to create a shared understanding of the problem space and inspire everyone to help solve the problem.

Visuals

Prepare a few slides introducing the problem and some compelling visuals or anecdotes to inspire empathy. The presentation should only take 10–15 minutes of your hour-long Creative Burst session.

Your introduction should include slides to share:

- Summary of the problem
- Personas involved
- Context of the problem within the personas' lives
- Quotes or video highlights from problem identification research
- How/why the problem isn't solved by your current product
- Potential value of the solution to your organization

Session activity

With the remaining 45–50 minutes, lead a discussion about the problem you've just shared. Chances are many of your team members will have individual experiences they can contribute.

Perhaps they've designed a solution to a tangential problem and can offer more detail about why this niche problem has been overlooked or deprioritized. Perhaps they've heard customers complain about the same problem during a research session and can share anecdotes that made an impression. Perhaps they even have a friend's or family member's personal experience they can relay.

Don't steer the discussion to solutions yet. Just encourage team members to share any insights to give everyone the most complete understanding of the problem you possibly can.

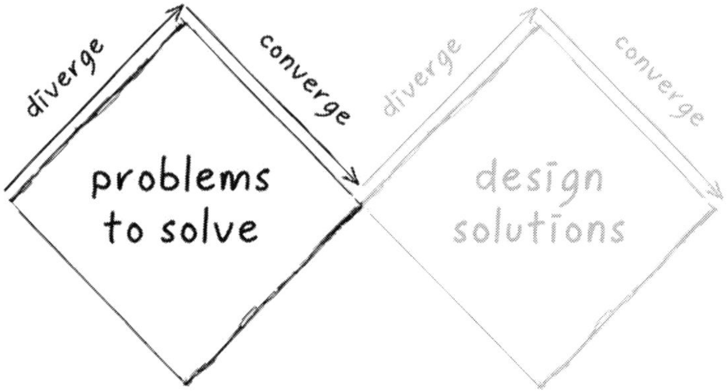

Follow-up exercise

The facilitator should take notes during the session, distill key takeaways, and share the summarized notes with everyone after the session.

For the group, the exercise isn't to start designing yet. Instead, task the team with finding inspiration in the world that relates to the problem. Looking at our example of task progress tracking, you might have team members find their favorite to-do list app, interesting treatments of progress bars, apps enabling different productivity methodologies, or even simply inspiring microinteractions for unique handling of checkboxes—whatever your team *feels* fits the problem space and inspires them to solve it within the context of your products.

Outcome

At least a few days before the next session, have participants each capture screenshots or brief screensharing videos for 2–3 of their favorite inspiration examples and save them to a specific location in a shared drive folder, or wherever your team captures visuals that aren't work output. Include team member names in filenames so you can ask questions if you need to.

EXAMPLE CREATIVE BURST: INSPIRATION SHARING

As you prepare for the second session, review all the inspiration images and videos. Clarify any that aren't self-explanatory with the team members who shared them. Make sure everything is in a single folder so you can page through them easily.

Make some notes about commonalities, trends, and particularly unique inspirations you find within the collection.

Goal

The goal of your inspiration sharing session is to inspire creativity and get team members thinking about interactions and visuals and technology beyond what they encounter during day-to-day projects. You want them thinking broadly when you move into your first divergent sketching session to help them get past the obvious solutions more quickly.

Visuals

Your primary visuals will be the inspiration images and videos. Keep the inspiration images simple—if you want, you can just share directly from the folder. Or create an inspirational project moodboard or collage using one of the many virtual whiteboard tools like Miro or Mural.

Then create a few slides to set the context for the day's activities.

Include:

- reminder of the problem
- trends from the inspirations
- expectations for the exercise

Session activity

Once you've shared those, spend your first 20 minutes stepping through the inspiration images and videos, encouraging the team to hold questions and comments until the end. Linger on each self-explanatory visual for several seconds to give your audience a chance to take it in. Linger a bit longer on more interesting ones or more complicated ones and invite the person who shared it to say a sentence or two explaining what it is before moving on.

When you've reached the end, open up discussion to the full group. Talk about trends you might have missed in the initial analysis. Review images people want to see again. Let people who didn't speak before share stories of why they chose specific products to share.

Most importantly, brainstorm how these inspirations relate to the problem you're collectively exploring. What opportunities did you see that you hadn't previously considered?

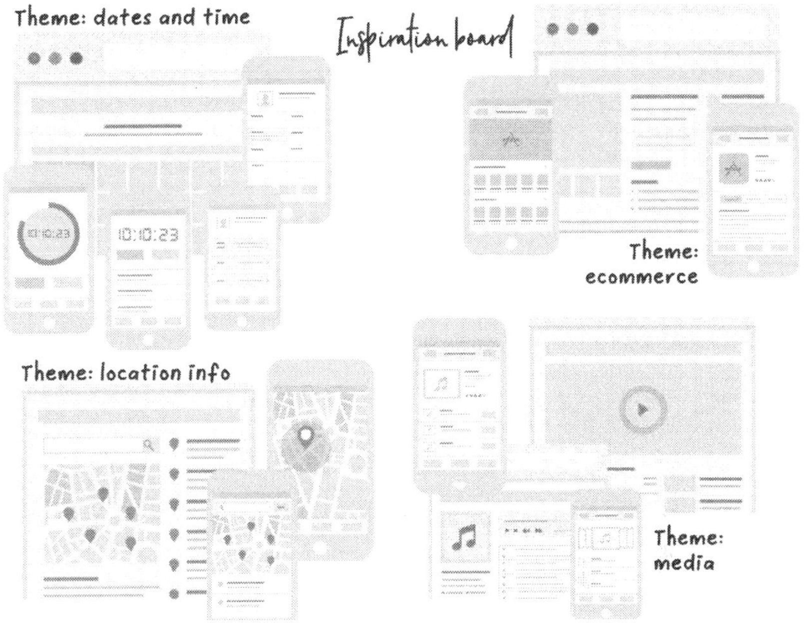

While you won't formally initiate sketching during the Creative Burst, encourage people to jot down notes and thumbnail sketches just to keep ideas fresh for the next phase.

Follow-up exercise

Task each participant with sketching several—I recommend 4–6—distinct design solutions, using the inspiration images and videos as creative idea starters.

If your team doesn't already have a culture of rapid sketching, set guidelines to keep people from getting overwhelmed by the prospect of creating so many distinct options.

Rough sketches, on paper. Single screen per design to showcase the essence of the concept. Maximum size of a half page to limit level of detail.

Outcome

Have each team member photograph their 4–6 designs and share all 4–6 images in a subfolder with their name inside a collective folder for the session several days before your next Creative Burst session. They should name each file with a number 1–6 and their own name.

EXAMPLE CREATIVE BURST: SOLUTION SHARING

Because the visuals come from the entire team, make sure all sketch photograph files open. If any files ended up outside folders named for individuals, create a folder for them. Create an empty "archive" folder inside each team member's folder, alongside the individual photos, and make sure each photo has a number 1–6 in the filename.

Create a folder alongside the team member folders called "reference designs" or something innocuous so you don't tip off designers that there will be a winnowing process. You don't want designers to turn a critical eye to their concepts at this stage.

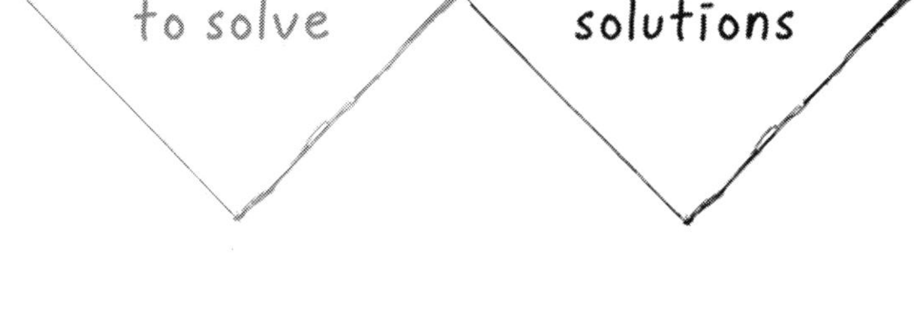

Lastly, create an online poll for people to choose from a list of numbers 1–6 that you can easily reset in between votes. You'll use this as a crude dot-voting mechanism for choosing the top designs after each team member shares their distinct concepts.

Goal

Your goal for the third full-group Creative Burst session is to fully embrace the divergent design process. You should have dozens of unique concepts from the previous session's "homework," and you'll use these as the basis for the next steps of deeper design exploration.

Visuals

You can create a slide with the day's agenda and rules for the activity if you like, but the photos of sketches you'll share with the group are the only visuals you really need.

Session activity

You will share the actual files to avoid the complication of swapping screens or forcing everyone to bring a laptop if you're meeting in person; but invite each designer to share his or her own work.

Give each person a few minutes to share all 4–6 designs, giving a sentence about the concept and any relevant inspiration. Then share the poll and have everyone vote for their favorite numbered concept. Leave the top two designs in the team member's folder and drag the remaining 2–4 designs into the archive folder. Copy the top two designs to the "reference designs" folder, but leave the originals where they are.

Reset the poll results and invite the next team member to share their concepts.

Once the group has voted for the strongest two concepts from each designer, do a final review of all designs that made it into the "reference designs" folder. Identify the most unique or compelling aspect of each, assign a keyword based on that, and rename the file to that keyword, removing the original number and person's name.

During my "tracking task progress" example from earlier, we might have chosen the keyword "pizza tracker" for a concept inspired by Domino's order tracking.

Follow-up exercise

Your team, driven by inspiration from the outside world, diverged to create dozens of initial designs. The group then identified the strongest ones and converged on a smaller set.

Ask which team members are interested in pursuing the problem space through the full solution design process. This is your core team. Assign them each to sketch 3–4 more designs, inspired by mashing up two of the design concepts in the "reference designs" folder. Allow team members to explore the problem in greater detail this time, so full-page designs are ok, but still limit each concept to 1–2 screens. From this point forward, you'll probably find yourselves falling into a more familiar design process, following your standard design collaboration and file sharing guidelines.

Outcome

While this won't result in as many concepts as you had at the beginning of this Creative Burst, creating 3–4 designs each that mash up two concepts you already identified as "strong" will have the multiplicative effect of another divergent design process.

Once everyone has completed their 3–4 designs, hold a smaller session with the core group to review all concepts, and ultimately choose how many distinct concepts, and which concepts, you want to move forward with.

Also, determine how each team member will contribute. Will

everyone own a concept? Will each team member contribute at key points, while a lead designer refines all concepts? These choices are yours to make.

If you're not comfortable eliminating promising ideas while they're still in such a rough sketched state, you can evolve your most promising 4–6 concepts until the designs reflect low-fidelity designs of a core task flow. But beware: the more designs you keep alive, the more work you do, and the more diluted team efforts will be. You'll ultimately pitch your single best solution, so the sooner you narrow your efforts to 2–3 of the absolute strongest concepts, the better the results will be.

CHAPTER FIFTEEN
YOUR LEARNING JOURNEY CONTINUES

THANK you for joining me on this journey through our Creative Burst process and beyond. I've shared a single case study of a single innovation program at a single company. I hope I've provided valuable learnings and takeaways and instilled you with the confidence—and a practical toolkit—to tackle innovation design within your own UX team.

If you want to learn more from others (which I hope you do), there are people and companies, many of whom have been doing this longer even than I've been a UX practitioner. Here are just a few perspectives and programs you might find valuable. I highly recommend you check them out.

OTHER INNOVATION PRACTICES

Google 20% time

Google famously encourages its software teams to devote 20% of their workweek[1] to personal projects that relate to Google's offerings. In fact, many of Google's most successful products owe their creation to such playful experiments—including the advertising revenue juggernaut AdSense and industry-leading free email service Gmail. While the program at Google has waxed and waned over the years—according to various internet articles—the core principle remains sound. Give your employees the freedom to pursue interesting projects with a portion of their workweek, and watch the grassroots innovation take hold.

3M

3M[2] is another organization famed for living up to the principle that promising ideas come from everywhere. Instead of relying solely on top-down strategic visions, 3M consistently listens to seemingly random ideas from its employees and creates market-leading products. In fact, a staple of UX practitioners everywhere—the humble Post-it note—was just one of the many employee-conceived products the company embraced.

Pixar's "Braintrust"

Computer-aided animation has become a fixture in our entertainment zeitgeist, and Pixar has been churning out movies with top-quality storytelling since many of us were kids. While those of us outside the animation industry might not appreciate the ongoing innovation of how they push boundaries with cutting-edge techniques in each new film, we can all learn something from how they accomplish it.

In his excellent book, *Creativity, Inc*[3], Pixar founder Ed Catmull describes the key role the Pixar "Braintrust" played in enabling their repeated innovation. A select committee of expert animators, producers, and storytellers review every project, guiding it through its fragile preliminary stages to successful completion. During the early stages, they offer insight, reassurance, and direction while teams struggle through shaping a project. During later stages, they act as brutally honest, yet constructive, gatekeepers of quality, helping teams reshape and polish projects until they truly shine.

RECOMMENDED RESOURCES

Remember how I mentioned that I was completely unqualified to write a generally applicable book about innovation? That is certainly not true of everyone. Here's a roundup of key resources that have inspired my process and impacted me most as a UX designer and advocate for innovation.

I've indicated my top picks, plus some others you might consider if you have more time—or get really into the topic.

Innovation

If you're ready to embrace the ambiguity of innovation as a concept and want to immerse yourself in a combination of inspiration and practical advice, I recommend starting with these books. In fact, if you only pick two books from all the resources I share, make it my two "Top Picks" for innovation. Prior to facilitating the program, these drove my passion for creating our Creative Bursts more than anything else.

- *Collective Genius*, Linda A. Hill, Greg Brandeau, Emily Truelove, and Kent Lineback [TOP PICK]
- *Creativity, Inc.*, Ed Catmull [TOP PICK]
- *The Innovator's Dilemma*, Clayton M. Christensen
- *Mapping Innovation*, Greg Satell
- *Disciplined Dreaming*, Josh Linkner

Creative work & UX design

I can't even begin to list all the books and thought leaders that have shaped my career as a UX designer and Content Strategist. But here are just a few that I find relevant to self-driven work like designer-led innovation projects. Some are methods, others share topics that many large software companies aren't exploring yet (as of this writing), and thus are areas designers can lead.

- *The Accidental Creative*, Todd Henry [TOP PICK]
- *Deep Work*, Cal Newport [TOP PICK]
- *The War of Art*, Steven Pressfield
- *The Paradox of Choice*, Barry Schwartz
- *The Jobs to Be Done Playbook*, Jim Kalbach
- *Figure It Out*, Stephen P. Anderson and Karl Fast

Facilitation

Whether you choose to act as the innovation leader for your UX team's innovation practice, volunteer to host an occasional Creative Burst yourself, or simply want to rally a smaller team at various points in the design process, you will need to run meetings. Meetings have developed a bad reputation in the business world. We have too many of them. They are a waste of time. But sometimes, we do need them. And they can be great.

These resources will help you improve your meetings so that you achieve the intended result, and no one leaves feeling like you've wasted their time.

- *Six Thinking Hats*, Edward de Bono [TOP PICK]
- *Gamestorming*, Dave Gray, Sunni Brown, and James Macanufo [TOP PICK]
- *Meeting Design*, Kevin M. Hoffman
- *Creative Confidence*, Tom Kelley and David Kelley
- *Flow*, Mihaly Csikszentmihalyi
- *Orchestrating Experiences*, Chris Risdon and Patrick Quattlebaum

Pitching

While there are plenty of excellent books that teach presentation design, my picks for helping you craft innovation design pitches aim to help with the hardest element to get right: constructing a compelling narrative. Between choosing the right words to convey your intent and learning how to break down the elements of your solution's story, these books will set you on the right track.

- *Words That Work*, Frank I. Luntz [TOP PICK]
- *Storynomics*, Robert McKee and Thomas Gerace [TOP PICK]
- *Presenting Design Work*, Donna Spencer
- *Back of the Napkin*, Dan Roam
- *Storytelling for User Experience*, Whitney Quesenbery and Kevin Brooks

I've limited each category to streamline this section and let you get on with the business of implementing what you've learned.

If you're interested in more book recommendations, either ones that simply didn't make the cut or I've learned about since publishing "Creative Burst," and perhaps some online tools to help you run your innovation program, you can find a more current list of resources online.

Plus learn about the companion workbook to guide you through creating your own innovation program.

www.creative-burst.com

PART FOUR

END MATTER

ACKNOWLEDGMENTS

I want to acknowledge special thanks to the following people, without whom this book would never have happened.

To my alpha readers, for their invaluable feedback. Your insights and critiques helped shape a simple story about my experiences into something that I hope will help others.

- Bob Thomas
- Chris Whelan
- Erin Fegely Kroberger
- Kirsten Robinson

To the fabulous team at Kronos who made Creative Bursts, and by extension this book, possible.

To my wonderful husband Jeff, whose years of living the dream of being a professional writer proved that books can happen if you just keep adding words to the page.

ABOUT THE AUTHOR

Kristin Morin began her professional web career in the early 2000s by dabbling with HTML and ended up calling herself a "web designer" back when no one truly knew what the term meant. After years in web and database development—and a bachelor's degree in Web Management in Internet Commerce from Johnson and Wales University—she found her true passion in User Experience (UX).

She graduated with a master's degree in Human Factors in Information Design from Bentley University, and in 2012 completed her professional transition from web development to UX. After six years working on websites as a user experience architect and content strategist in web agencies, she made the switch to staff product designer at an enterprise software company: Kronos.

She prides herself on her dedication to lifelong learning, constantly seeking to expose herself to thought leadership to learn to be a better designer than she was yesterday. Her professional areas of focus include information visualization, UX writing, content strategy, design strategy, encouraging cross-functional collaboration, and—of course—facilitating creativity and innovation.

CONNECT WITH KRISTIN MORIN

- Web: www.kristinmorin.com
- LinkedIn: linkedin.com/in/kristinmorin/

NOTES

1. INNOVATION WHO, WHAT, WHY, AND HUH?

1. Clayton M. Christensen, Michael E. Raynor, and Rory McDonald, "What Is Disruptive Innovation?", *Harvard Business Review*, December 2015, accessed February 9, 2022, https://hbr.org/2015/12/what-is-disruptive-innovation
2. Clayton M. Christensen, *The Innovator's Dilemma* (Boston: Harvard Business Review Press, 2015)

9. OUTGROWING CREATIVE BURSTS

1. Therese Fessenden, "The Diverge-and-Converge Technique for UX Workshops," *Nielsen Norman Group (blog)*, December 15, 2019, accessed February 9, 2022, https://www.nngroup.com/articles/diverge-converge/

11. BUILDING BLOCK 1—PROBLEM IDENTIFICATION

1. Barry Schwartz, *The Paradox of Choice* (New York: HarperCollins, 2009)

12. BUILDING BLOCK 2—SOLUTION DESIGN

1. Five Sketches Consulting Inc., "Use The Five Sketches™ Process to Get Quality Software Designs," *Five Sketches Consulting*, accessed February 9, 2022, https://fivesketches.com/quality-software-designs-by-sketching/

15. YOUR LEARNING JOURNEY CONTINUES

1. Bill Murphy Jr., "Google Says It Still Uses the '20-Percent Rule,' and You Should Totally Copy It," *Inc.com*, November 1, 2020, accessed February 9, 2022, https://www.inc.com/bill-murphy-jr/google-says-it-still-uses-20-percent-rule-you-should-totally-copy-it.html
2. "The Strategy That Makes 3M an Innovation Powerhouse" by Suryakant Trivedi, published May 27, 2021 on The Strategy Story, accessed February 9, 2022. Access at https://thestrategystory.com/2021/05/27/3m-innovation-strategy/

   Suryakant Trivedi, "The Strategy That Makes 3M an Innovation

Powerhouse," *The Strategy Story*, May 27, 2021, accessed February 9, 2022, https://thestrategystory.com/2021/05/27/3m-innovation-strategy/

3. *Creativity, Inc.* by Ed Catmull, April 8, 2014 by Random House
   Ed Catmull, *Creativity, Inc.* (New York: Random House, 2014)